EIT
Civil
Review

Second Edition

Dr. Donald G. Newnan, P.E.
Editor

Contributors:
Dr. Alan Williams, S.E., C.Eng.
California Department of Transportation
Dr. Bruce Larock, P.E.
University of California, Davis
Dr. Kenneth Williamson, P.E.
Oregon State University
Dr. Braja Das, P.E.
California State University, Sacramento
Dr. Robert W. Stokes
Kansas State University
Dr. Michael Taylor
University of California, Davis
Lincoln D. Jones, P.E.
San Jose State University

Engineering Press　　　　　　　　　　Austin, Texas

ISBN 1-57645-013-9
5 4 3 2 1

Engineering Press
P.O. Box 200129
Austin, TX 78720-0129

(800) 800-1651
FAX (800) 700-1651

Contents

After a 30-year career in the environmental engineering area, I served 10 years as the Director of Examinations for the National Council of Examiners for Engineering and Surveying (NCEES). The NCEES develops and grades the examinations that all 50 states and 5 jurisdictions of the USA use in licensing engineers and land surveyors. During my time at the NCEES, approximately 500,000 candidates sat for the Fundamentals (FE/EIT) and about 300,000 took the Principles and Practice (PE) examinations.

Engineering Press has provided top quality products over a period of years with guidance to candidates who are preparing to take the exams. Their publications are kept up-to-date with the many changes that have been made in the exams without overloading them with extraneous information.

The specifications for all the exams are developed by surveys of the practicing professional engineers and the education professionals who teach in the accredited engineering programs at universities and colleges throughout the United States. Item (problem, question) writers from all over the country participate in developing the items that are used on all exams. They must be licensed in at least one of the 55 states or jurisdictions and subject matter experts in a particular area. They come from all geographical areas and represent different areas of practice such as consulting, private practice, education and government. Many of them serve on the various state licensing boards.

The actual exams are assembled by committee members with the same representation as above. Committee members are also expected to write items. It is estimated that each item on a particular test will have been reviewed by at least 5 - 10 individuals over a one to one and a half year period of time before it appears on an examination. All reviews are conducted to ensure the items have only one correct answer (multiple-choice), are clear, unambiguous and of the appropriate difficulty level. In spite of this careful process, flawed items are occasionally found on the tests.

During scoring process and before any results are sent to the state boards, various statistical methods are used to provide a quality check on the performance of each item on the test. Sometimes seriously flawed items may be eliminated from scoring or other corrections made. In other words, credit may be given to all candidates regardless of their response. All decisions are made by subject matter experts with the best interest of the candidates, consistent with public protection, in mind.

I encourage you to take advantage of the publications provided by Engineering Press for success on the first attempt. If you have specific questions, use the web site at: http://www.engrpress.com. For detailed information on the exams refer to the NCEES web site at: http://www.ncees.com.

J. Earl Herndon Jr., P. E., DEE
Former Director of Exams

Preface

The Fundamentals of Engineering/Engineer-In-Training (FE/EIT) exam has dramatically changed. For about the last fifty years the 8-hour exam has been a test of a dozen topics that are the core of every engineer's education: Math-Chem-Physics-Statics-Dynamics-Fluids-Thermo-Circuits-Engr Econ-Mat Sci-Computers and Mech of Materials. All that changed in the Fall of 1996. Now the exam has two distinct parts to it. The morning four-hour portion is pretty much like the old exam with about the same dozen topics.

The afternoon four-hour portion of the exam is new and different. In fact there are six different afternoon exams. Each one covers a different branch of engineering: civil, mechanical, electrical, chemical, industrial and general. While the morning exam covers the first 2-2½ years of an engineer's education, the afternoon exam covers the last 1½-2 years of an engineers education in his specific branch of engineering. This book has been revised in this edition to closely follow the afternoon topics and the afternoon exam content of the *Fundamentals of Engineering Reference Handbook*. This means that an orderly review in preparation for the exam has two components and requires two books. Everyone can use the *Engineer-In-Training License Review* book for the morning exam, as can general engineers for the afternoon exam. All other engineers will need one of the following:

EIT Civil Review   *EIT Chemical Review*
EIT Mechanical Review   *EIT Industrial Review*
EIT Electrical Review

To prepare this book I assembled a carefully selected team of professors, with each professor preparing that portion of the book that relates to his expertise. I am most grateful to Professors Williams, Larock, Williamson, Das, Taylor, Jones and Stokes for devoting a lot of time and effort to writing this book. If you encounter any errors in this book, we would appreciate being notified. Notes and comments may be sent directly to me, c/o Engineering Press, P.O. Box 200129, Austin, TX 78720-0129.

These books, in their previous editions, have helped tens of thousands of engineers prepare for and pass the FE/EIT exam. The team of engineering professors and I have worked diligently to provide you the best possible review to quickly and efficiently prepare for this changed exam. We wish you good luck in the exam.

Donald G. Newnan

Selected books of interest published by Engineering Press
LICENSE REVIEW BOOKS
For the FE/EIT Examination--2 Books are needed:
Engineer-In-Training License Review
and One of these supplements:
EIT Civil Review
EIT Mechanical Review
EIT Electrical Review
EIT Chemical Review
EIT Industrial Review
For the PE Examinations:
Civil Engineering License Review
Civil Engineering License Problems and Solutions
Environmental Engineering Problems and Solutions
Seismic Design of Buildings and Bridges
Mechanical Engineering License Review
Electrical Engineering License Review
Electrical Engineering License Problems and Solutions
A Programmed Review for Electrical Engineering
Chemical Engineering License Review
Structural Engineer Registration: Problems and Solutions

EXAM FILES

Calculus I   *Materials Science*
Calculus II   *Mechanics of Materials*
Calculus III   *Organic Chemistry*
College Algebra   *Physics I —Mechanics*
Circuit Analysis   *Physics III —Electricity and Magnetism*
Differential Equations   *Probability and Statistics*
Dynamics   *Statics*
Engineering Economic Analysis   *Thermodynamics*
Fluid Mechanics
Linear Algebra

OTHER BOOKS

Being Successful As An Engineer   *Engineering Economic Analysis*
The Basics of Structural Analysis   *Elements of Bioenvironmental Engineering*
Fluid Mechanics   *Soil Mechanics Laboratory Manual*

Free Catalog
For a free catalog describing the books above and 80 other license review books,
with prices and ordering information,
write to:
Engineering Press Bookstore   FAX: (800) 700-1651
P.O. Box 200129   Phone: (800) 800-1651
Austin, TX 78720-0129

INTRODUCTION

BECOMING A PROFESSIONAL ENGINEER

To achieve registration as a Professional Engineer, there are four distinct steps: 1) education, 2) the Fundamentals of Engineering/Engineer-In-Training (FE/EIT) exam, 3) professional experience, and 4) the professional engineer exam. These steps are described in the following sections.

Education
Generally, no college degree is required to be eligible to take the FE/EIT exam. The exact rules vary, but all states allow engineering students to take the FE/EIT exam before they graduate, usually in their senior year. Some states, in fact, have no education requirement at all. One merely need apply and pay the application fee. Perhaps the best time to take the exam is immediately following completion of related course work. For most engineering students, this will be in the senior year.

Fundamentals of Engineering/
Engineer-In-Training Examination
This eight-hour, multiple-choice examination is known by a variety of names: Fundamentals of Engineering, Engineer-In-Training (EIT), or Intern Engineer, but no matter what it is called, the exam is the same in all states. It is prepared and graded by the National Council of Examiners for Engineering and Surveying (NCEES).

Experience
States that allow engineering seniors to take the FE/EIT exam have no experience requirement. These same states, however, generally will allow other applicants to substitute acceptable experience for course work. Still other states may allow one to take the FE/EIT exam without any education or experience requirements.

Professional Engineer Examination
The second national exam is called Principles and Practice of Engineering by NCEES, but many refer to it as the Professional Engineer exam or P.E. exam. All states, plus Guam, the District of Columbia, and Puerto Rico use the same NCEES exam. Typically, four years of acceptable experience is required before one can take the Professional Engineer exam, but the requirement may vary from state to state. Review materials for this exam are found in other engineering license review books.

FUNDAMENTALS OF ENGINEERING/
ENGINEER-IN-TRAINING EXAMINATION

Background
Laws have been passed that regulate the practice of engineering to protect the public from incompetent practitioners. Beginning about 1907 the individual states began passing *title* acts regulating who could call themselves an engineer. As the laws were strengthened, the *practice* of engineering was limited to those who were registered engineers, or to those

working under the supervision of a registered engineer. Originally the laws were limited to civil engineering, but over time they have evolved so that the *titles* and sometimes the *practice* of most branches of engineering are included. There is no national registration law; registration is based on individual state laws and is administered by boards of registration in each state.

Examination Development

Initially, the states wrote their own examinations, but beginning in 1966 the NCEES took over the task for some of the states. Presently the NCEES exams are used by all states. Now it is easy for engineers that move from one state to another to achieve registration in the new state. About 50,000 engineers take the FE/EIT exam annually. This equals about half the engineers graduated in the U.S. each year.

The development of the FE/EIT exam is the responsibility of the NCEES Committee on Examinations for Professional Engineers. The committee is composed of people from industry, consulting, and education, some of whom are subject-matter experts. The test is intended to evaluate an individual's understanding of basic sciences as well as engineering sciences along with one's ability to apply this knowledge to engineering problems. Every five years or so, NCEES conducts an engineering task analysis survey. People in industry, consulting, and education are surveyed to review engineering practice and the knowledge needed for it. From this, NCEES develops what they call a "matrix of knowledge" that forms the basis for the FE/EIT exam structure described in the next section.

The actual exam questions are prepared by the NCEES committee members, subject matter experts, and other volunteers. All people participating must hold professional registration. When the questions have been written, they are circulated for review in workshop meetings and by mail. Both single- and multi-part questions are used; the latter are arranged so the solution of each succeeding part does not depend on the correct solution of a previous part. All problems are four-way multiple choice.

Examination Structure

The FE/EIT exam is divided into a morning four-hour section and an afternoon four-hour section. The Fundamentals of Engineering Examination in the afternoon period has changed to a discipline specific examination. Six exams are presented in the afternoon test booklet, one for each of the following five disciplines: Civil, Mechanical, Electrical, Chemical, Industrial; and a General topic for those examinees not covered by the five engineering disciplines. The test consists of 60 problems in the topic. This may be of great benefit to those specializing in one of the five major engineering disciplines. Most of the test's topics cover the third and fourth year of your college courses. These are the courses which you will use for the balance of your engineering career, so the test becomes focused to your own needs. Graduate engineers will find the PM discipline test to their advantage as the broad fundamentals test usually causes many engineers to do a good deal of review of the earliest classwork.

The major subjects for the afternoon sections are:

Civil	No.Probs.
Soil Mechanics & foundations	6
Structural Analysis, frames, trusses	6
Hydraulics & hydro systems	6
Structural Design, concrete, steel	6
Environmental Engineering, waste water, solid waste treatment	6
Transportation facilities, highways railways, airports	6
Water purification and treatment	6
Surveying	6
Computer and Numerical methods	6
Legal & professional aspects	3
Construction Management	3

Mechanical	
Mechanical Design	6
Dynamic Systems,Vibration,Kinematics	6
Thermodynamics	6
Heat Transfer	6
Fluid Mechanics	6
Stress Analysis	6
Measurement and Instrumentation	6
Material Behavior/ Processing	3
Computer, Automation, Robotics and Numerical methods	3
Energy Conversion and Power Plants	3
Automatic control	3
Refrigeration and HVAC	3
Fans, Pumps, and Compressors	3

Electrical	
Digital Systems	6
Analog Electronic Circuits	6
Electromagnetic Theory & Applications	6
Network Analysis	6
Control Systems Theory and Analysis	6
Solid State Electronics and Devices	6
Communications Theory	6
Signal Processing	3
Power Systems	3
Computer Hardware Engineering	3
Computer Software Engineering	3
Instrumentation	3
Computer and Numerical Methods	3

Chemical	No.Probs
Material & Energy Balances	9
Chemical Thermodynamics	6
Mass Transfer	6
Chemical Reaction Engineering	6
Process Design and Econ Evaluation	6
Heat Transfer	6
Transport Phenomenon	6
Process Control	3
Process Equipment Design	3
Computer and Numerical Methods	3
Process Safety	3
Pollution Prevention	3

Industrial	
Production Planning & Scheduling	3
Engineering Economics	3
Engineering Statistics	3
Statistical Quality Control	3
Manufacturing Processes	3
Mathematical Optimization-Modeling	3
Simulation	3
Facility design and location	3
Work Performance and Methods	3
Manufacturing Systems Design	3
Industrial Ergonomics	3
Industrial Cost Analysis	3
Material Handling System Design	3
Total Quality Management	3
Computer Computations-Modeling	3
Queuing Theory and Modeling	3
Design of Industrial Experiments	3
Industrial Management	3
Information System Design	3
Productivity Measurement	3

General	
Mathematics	12
Electrical Circuits	6
Statics	6
Thermodynamics	6
Chemistry	4.5
Dynamics	4.5
Fluid Mechanics	4.5
Mechanics of Materials	4.5
Material Science	3
Computers	3
Ethics	3
Engineering Econ	3

Taking the Examination
Examination Dates
The National Council of Examiners for Engineering and Surveying (NCEES) prepares FE/EIT exams for use on a Saturday in April and October each year. Some state boards administer the exam twice a year; others offer the exam only once a year. The scheduled exam dates are:

	April	October
1998	25	31
1999	24	30
2000	15	28

Those wishing to take the exam must apply to their state board several months before the exam date. We recommend that Civil, Mechanical, Electrical, Chemical, and Industrial engineers take the discipline specific exam. All others should take the General examination.

Examination Procedure
Before the morning four-hour session begins, the proctors will pass out exam booklets and a scoring sheet to each examinee. An HB or #2 pencil is to be used to record answers. Space is provided on each page of the examination booklet for scratchwork. The scratchwork will *not* be considered in the scoring.

The examination is closed book. You may not bring any reference materials with you to the exam. To replace your own materials, NCEES has prepared a *Fundamentals of Engineering (FE) Reference Handbook*. The handbook contains engineering, scientific and mathematical formulas and tables for use in the examination. Examinees will receive the handbook from their state registration board prior to the examination. The *FE Reference Handbook* also is included in the exam materials distributed at the beginning of each four-hour exam period. If you do not already have a copy of the *FE Reference Handbook* you should purchase one right away.

There are three versions (A, B, and C) of the exam. These have the major subjects presented in a different order to reduce the possibility of examinees copying from one another. The first subject on your exam, for example, might be fluid mechanics, while the exam of the person next to you may have electrical circuits as the first subject.

The afternoon session begins following a one-hour lunch break. The afternoon exam booklets will be distributed along with a scoring sheet. At the beginning of the afternoon test period, the examinee will mark the answer sheet as to which exam he is taking. There will be 60 multiple choice questions, each of which carries twice the grading weight of the morning exam questions.

If you answer all questions more than 30 minutes early, you may turn in the exam materials and leave. In the last 30 minutes, however, you must remain to the end of the exam period to ensure a quiet environment for all those still working, and to ensure an orderly collection of materials.

Examination-Taking Suggestions
Those familiar with the psychology of examinations have several suggestions for examinees:

1. There are really two skills that examinees can develop and sharpen. One is the skill of illustrating one's knowledge. The other is the skill of familiarization with examination structure and procedure. The first can be enhanced by a systematic review of the subject matter. The second, exam-taking skills, can be improved by practice with sample problems, that is, problems that are presented in the exam format with similar content and level of difficulty.

2. Examinees should answer every problem, even if it is necessary to guess. There is no penalty for guessing. The best approach to guessing is to first eliminate the one or two obviously incorrect answers among the four alternatives. If this can be done, the chance of selecting a correct answer obviously improves from 1 in 4 to 1 in 2 or 3.

3. Plan ahead with a strategy and a time allocation. There are 120 morning problems in 12 major subject areas. Compute how much time you will allow for each of the 12 subject areas. You might allocate a little less time per problem for those areas in which you are most proficient, leaving a little more time in subjects that are more difficult for you. Your time plan should include a reserve block for especially difficult problems, for checking your scoring sheet, and finally to make last-minute guesses on problems you did not work. A time plan gives you the confidence of being in control. Misallocation of time for the exam can be a serious mistake. Your strategy might also include time allotments for two passes through the exam: the first to work all problems for which answers are obvious to you; the second to return to the more complex, time-consuming problems and the ones at which you might need to guess.

4. Read all four multiple choice answers before making a selection. The first answer in a multiple choice question is sometimes a plausible decoy-not the best answer.

5. Do not change an answer unless you are absolutely certain you have made a mistake. Your first reaction is likely to be correct.

6. If time permits, check your work.

7. Do not sit next to a friend, a window, or other potential distraction.

License Review Books

There are two rules that we suggest you follow in selecting license review books to ensure that you obtain up-to-date materials:

1. Consider only the purchase of materials that have a 1993 or later copyright. Prior to October, 1993 reference books could be brought to the exam, and some review books added unnecessary reference materials to get them accepted as reference books. That strategy was successful then, but it makes no sense now. Nothing can be taken into the exam, and the unnecessary material in these older books diverts attention from the exam content. Look for books that cover what you need to know--no more, and no less.

2. Even if a license review book has a recent copyright date, is the content up to date? Books that ignore the multiple-choice format, or include irrelevant material, are not up to date.

Textbooks

If you still have your university textbooks, they can be useful in preparing for the exam, unless they are too out of date. To a great extent the books will be like old friends with familiar notation. You probably need both textbooks and license review books for efficient study and review. You must, however, pay close attention to the *FE Reference Handbook and* the notation used in it, because it is the only book you will have in the exam.

Examination Day Preparations

There is no doubt that the exam day will be a stressful and tiring one. You should take steps to eliminate the possibility of unpleasant surprises. If at all possible, visit the examination site ahead of time. Try to determine such items as:

1. How much time should I allow for travel to the exam on that day. Plan to arrive about 15 minutes early. That way you will have ample time, but not too much time. Arriving too early and mingling with others who also are anxious, can increase your anxiety and nervousness.

2. Where will I park?

3. How does the exam site look. Will I have ample work space? Will it be overly bright (sunglasses) or cold (sweater), or noisy (earplugs)? Would a cushion make the chair more comfortable?

4. Where is the drinking fountain? Lavatory facilities? Pay phone?

5. What about food? Should I take something, along for energy in the exam? A light bag lunch during the break probably makes sense.

Items to Take to the Examination

● Calculator-NCEES says you may bring a battery-operated, silent, non-printing calculator. You need to determine whether or not your state permits pre-programmed calculators. Bring extra batteries for your calculator just in case; and many people feel that bringing a second calculator is also a very good idea.

● Clock-You must have a time plan and a clock or wristwatch.

● Pencils-You should consider using mechanical pencils so you don't have to run around sharpening pencils. And you surely do not want to drag along a pencil sharpener.

● Eraser-Try a couple to decide what to bring along. You must be able to change answers on the multiple choice answer sheet, and that means a good eraser.

● Exam Assignment Paperwork-Take along the letter assigning you to the exam at the specified location to prove that you are the registered person. Also bring something with your name and picture (driver's license or identification card).

● Items Suggested By Your Advance Visit--If you visit the exam site, it probably will suggest an item or two that you need to add to your list.

● Clothes-Plan to wear comfortable clothes. You probably will do better if you are slightly cool, so it is wise to wear layered clothing.

Examination Scoring

The questions are machine-scored by scanning. The answer sheets are checked for errors by computer. Marking two answers to a question, for example, will be detected and no credit given.

Passing the FE/EIT Examination

The 120 morning and 60 double-weighted afternoon problems yield a raw score of 240. The passing score is set as the equivalent of a raw score of 124 out of 280 on the October, 1990 examination. To translate this established level of difficulty to new examinations, a portion of each exam contains problems given on previous exams. An analysis of this subset of problems can be used to scale the new exam raw score to the same level of difficulty as 124 on the October, 1990 exam. This produces a standardized exam where the percentage of applicants passing, may vary from exam to exam, but they all demonstrate an established level of competency. Since most state licensing laws require 70 on the exam to pass, the raw passing score (equivalent to 124 on the October, 1990 exam) is converted to 70 and reported to the states accordingly. Higher raw scores are reported proportionally higher than 70; lower raw scores as proportionally lower than 70. What does it all mean? You are faced with 120 single-weighted morning problems and 60 double-weighted afternoon problems which makes a maximum possible score of 240. You must correctly answer problems with an equivalent score of about 124 to receive a passing score. So, you do *not* need to answer 70% of the problems correctly; you will get a converted score of 70 if you answer about 45% of the problems correctly.

The converted scores are reported to the individual state boards in about two months, along with the recommended pass or fail status of each applicant. The state board is the final authority of whether an applicant has passed or failed the exam.

Although there is some variation from exam to exam, the following gives the approximate passing rates:

Applicant's Degree	Percent Passing Exam %
Engineering from accredited school	62
Engineering from non-accredited school	50
Engineering Technology from accredited school	42
Engineering Technology from non-accredited school	33
Non-Graduates	36
First-time Applicants	67
All Applicants	56

If you do not pass the exam on your first attempt, you can re-apply and take it again.

Chapter 1
Geotechnical
Dr. Braja M. Das

Particle Size Distribution

The particle size distribution in a given soil is determined in the laboratory by sieve analysis and hydrometer analysis. For classification purposes in coarse-grained soils, the following two parameters can be obtained from a particle size distribution curve:

$$\text{Uniformity coefficient, } c_u = \frac{D_{60}}{D_{10}} \qquad (1)$$

$$\text{Coefficient of gradation, } c_c = \frac{D_{30}^2}{D_{60} \times D_{10}} \qquad (2)$$

where D_{10}, D_{30}, D_{60} = diameters through which, respectively, 10%, 30%, and 60% soil pass.

Weight-Volume Relationships

Soils are three-phase systems containing soil solids, water, and air (Fig. 1). Referring to Fig. 1

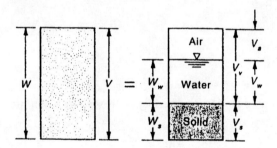

Figure 1

$$W = W_s + W_w \qquad (3)$$

$$V = V_s + V_v = V_s + V_w + V_a \qquad (4)$$

where W = total weight of the soil specimen, W_s = weight of the solids, W_w = weight of water, V = total volume of the soil, V_s = volume of soil solids, V_v = volume of voids, V_w = volume of water, and V_a = volume of air.

The *volume relationships* can then be given as follows:

$$\text{Void ratio } = e = \frac{V_v}{V_s} \qquad (5)$$

$$\text{Porosity } = n = \frac{V_v}{V} = \frac{e}{1+e} \qquad (6)$$

$$\text{Degree of saturation } = S(\%) = \frac{V_w}{V_v} \times 100 = \frac{wG}{e} \times 100 \tag{7}$$

Similarly, the *weight relationships* are:

$$\text{Water content (or moisture content) (\%) } = w = \frac{W_w}{W_s} \times 100 \tag{8}$$

$$\text{Moist unit weight } = \gamma = \frac{W}{V} = \frac{G\gamma_w(1+w)}{1+e} \tag{9}$$

$$\text{Dry unit weight } = \gamma_d = \frac{W_s}{V} = \frac{\gamma}{1+w} = \frac{G\gamma_w}{1+e} \tag{10}$$

$$\text{Unit weight of solids } = \gamma_s = \frac{W_s}{V_s} = G\gamma_w \tag{11}$$

In the preceding equations, γ_w = unit weight of water (62.4 lb/ft^3 or 9.81 kN/m^3), and G = specific gravity of soil solids, or

$$G = \frac{W_s}{V_s\gamma_w} \tag{12}$$

Relative Density

In granular soils, the degree of compactioñ is generally expressed by a nondimensional parameter called *relative density*, D_d, or

$$D_d = \frac{e_{max} - e}{e_{max} - e_{min}} \tag{13}$$

where e = actual void ratio in the field, e_{max} = void ratio in the loosest state, and e_{min} = void ratio in the densest state.

In terms of dry unit weight

$$D_d = \frac{\dfrac{1}{\gamma_{min}} - \dfrac{1}{\gamma_d}}{\dfrac{1}{\gamma_{min}} - \dfrac{1}{\gamma_{max}}} \tag{14}$$

where γ_{min}, γ_{max} = minimum and maximum *dry* unit weights, respectively; and γ_d = dry unit weight in the field.

Consistency of Clayey Soils

The moisture content in percent at which the cohesive soil will pass from a liquid state to a plastic state is called the **liquid limit**. Similarly, the moisture contents at which the soil changes from a plastic state to a semisolid state, and from a semisolid state to a solid state are referred to as the **plastic limit** and the **shrinkage limit**, respectively. These limits are referred to as the **Atterberg limits** (Fig. 2).

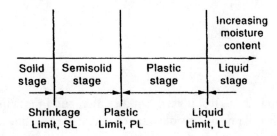

Figure 2.

Liquid limit (LL) is the moisture content in percent at which the groove in the Casagrande liquid limit **device** closes for a distance of 0.5 in. after 25 blows.

Plastic limit (PL) is the moisture content in percent at which the soil, when rolled into a thread of 1/8-in. in diameter crumbles.

Plasticity index (PI) is defined as

$$PI = LL - PL \qquad (15)$$

Shrinkage limit (SL) is the moisture content at which the volume of the soil mass no longer changes.

Shrinkage index (SI) is defined as

$$SI = PL - SL \qquad (16)$$

Permeability

The rate of flow of water through a soil of gross cross-sectional area, A, can be given by the following relationships (Fig. 3) called Darcy's law

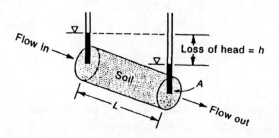

Figure 3.

$$v = ki \qquad (17)$$

where v = discharge velocity, k = coefficient of permeability, I = hydraulic gradient = h/L (see Fig. 3).

$$Q = vA = kiA \tag{18}$$

where Q = flow through soil in unit time and A = area of cross section of the soil at a right angle to the direction of flow. Or

$$k = \frac{Q}{iA} \tag{19}$$

Flow Nets

In many cases, flow of water through soil varies in direction and in magnitude over the cross section. In those cases, calculation of rate of flow of water can be made by using a graph called a **flow net**. A flow net is a combination of a number of flow lines and equipotential lines. A flow line is one along which a water particle will travel from the upstream to the downstream side. An equipotential line is one along which the potential head at all points is the same.

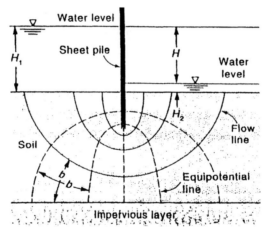

Figure 4.

Figure 4 shows an example of a flow net in which water flows from the upstream to the downstream around a sheet pile. Note that in a flow net the flow lines and equipotential lines cross at *right angles*. Also the flow elements constructed are *approximately square*. Referring to Fig. 4, the flow line in unit time (Q) per unit length normal to the cross section shown is

$$Q = k\frac{N_f}{N_d} H \tag{20}$$

where N_f = number of flow channels, N_d = number of drops, and H = head difference between the upstream and downstream side. (Note: In Fig. 4, N_f = 4 and N_d = 6).

Consolidation

Consolidation settlement is the result of volume change in saturated clayey soils due to the expulsion of water occupied in the void spaces. In soft clays, the major portion of the settlement of a foundation may be due to consolidation. Based on the theory of consolidation, a soil may be divided into two major categories: (a) normally consolidated and (b) over consolidated. For *normally consolidated clay*, the *present effective overburden pressure* is the maximum to which the soil has been subjected in the recent geologic past.

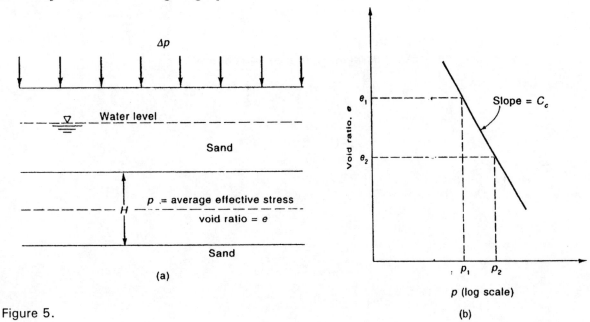

Figure 5.

(a)

(b)

Figure 5(a) shows a normally consolidated clay deposit of thickness H. The consolidation settlement (ΔH) due to a surcharge Δp can be determined as:

$$\Delta H = \frac{C_c H}{1 + e_i} \log\left(\frac{p_i + \Delta p}{p_i}\right) \tag{21}$$

where e_i = initial void ratio, p_i = initial *average* effective pressure, and C_c = compression index.
The compression index can be determined from the laboratory as [Fig. 5(b)].

$$C_c = \frac{e_1 - e_2}{\log\left(\dfrac{p_2}{p_1}\right)} \tag{22}$$

Shear Strength

The shear strength of a soil (s), in general, is given by the **Mohr-Coulomb failure criteria**, or

$$s = c + \sigma' \tan\phi \tag{23}$$

where c = cohesion, σ' = effective normal stress, and ϕ = drained friction angle. For sands, $c = 0$, and the magnitude of ϕ varies with the relative density of compaction, size, and shape of the soil particles.

For *normally consolidated clays*, $c = 0$. So $s = \sigma'$ tan ϕ. However, for over consolidated clays $c \neq 0$, thus $s = c + \sigma'$ tan ϕ. An important concept for the shear strength of cohesive soils is the so-called $\phi = 0$ concept. This is the condition where drainage from the soil does not take place during loading. For such a case

$$s = c_u \tag{24}$$

where c_u is the undrained shear strength.

The *unconfined compression strength*, q_u, of a cohesive soil is

$$q_u = 2c_u \tag{25}$$

Problems

1. A moist soil specimen has a volume of 0.15 m^3 and weighs 2.83 kN. The water content is 12% and the specific gravity of soil solids is 2.69. Determine:
    a. Moist unit weight, γ
    b. Dry unit weight, γ_d
    c. Void ratio, e
    d. Degree of saturation, S

2. For a soil deposit in the field, the dry unit weight is 14.9 kN/m^3. From the laboratory, the following were determined: $G = 2.66$, $e_{max} = 0.89$, $e_{min} = 0.48$. Find the relative density in the field.

3. For a sandy soil, the maximum and minimum void ratios are 0.85 and 0.48 respectively. In the field, the relative density of compaction of the soil is 29.3%. Given $G = 2.65$. Determine the moist unit weight of the soil at $w = 10\%$.

4. Refer to the flow net shown in Fig. 4. Given: $k = 0.03$ cm/min, $H_1 = 10$ min, and $H_2 = 1.8$ m. Determine the seepage loss per day per foot under the sheet pile construction.

5. The results of a sieve analysis of a granular soil are as follow:

U.S. Sieve No.	Sieve Opening (mm)	Percent Retained on Each Sieve
4	4.75	0
10	2.00	20
40	0.425	20
60	0.25	30
100	0.15	20
200	0.075	5

    Determine the uniformity coefficient and coefficient of gradation of the soil.

6. For a normally consolidated clay of 3.2 m thickness, the following are given:

    Average effective pressure = 98 kN/m^2
    Initial void ratio = 1.1
    Average increase of pressure in the clay layer = 42 kN/m^2
    Compression index = 0.27

    Estimate the consolidation settlement

7. An oedometer test in a normally consolidated clay gave the following results.

Average effective pressure (kN/m^2)	Void ratio
100	0.9
200	0.82

Calculate the compression index.

Solutions

Solution 1.

Part a.

$$\gamma = \frac{W}{V} = \frac{2.83}{0.15} = 18.87 \text{ kN} / \text{m}^3$$

Part b:

$$\gamma_d = \frac{\gamma}{1+w} = \frac{18.87}{1+\left(\dfrac{12}{100}\right)} = 16.85 \text{ kN} / \text{m}^3$$

Part c:

$$\gamma_d = \frac{G\gamma_w}{1+e}; \quad e = \frac{G\gamma_w}{\gamma_d} - 1 = \frac{(2.69)(9.81)}{16.85} - 1 = 0.566$$

Part d:

$$S = \frac{wG}{e} \times 100 = \frac{(0.12)(2.69)}{0.566} \times 100 = 57.03\%$$

Solution 2. In the field

$$\gamma_d = \frac{G\gamma_w}{1+e}; \quad e = \frac{G\gamma_w}{\gamma_d} - 1 = \frac{(2.66)(9.81)}{14.9} - 1 = 0.75$$

$$D_d = \frac{e_{max} - e}{e_{max} - e_{min}} = \frac{0.85 - 0.75}{0.89 - 0.48} = 34\%$$

Solution 3.

$$D_d = 0.293 = \frac{e_{max} - e}{e_{max} - e_{min}} = \frac{0.85 - e}{0.85 - 0.45}; \quad e = 0.733$$

$$\gamma = \frac{G\gamma_w(1+w)}{1+e} = \frac{(2.65)(9.81)(1+0.1)}{1+0.733} = 16.5 \text{ kN} / \text{m}^3$$

Solution 4.

$$Q = k\frac{N_f}{N_d}H = \frac{(0.03 \times 60 \times 24 \text{ cm}/\text{day})}{100}\left(\frac{4}{6}\right)(10-1.8) = 2.36 \text{ m}^3/\text{day}/\text{m}$$

Solution 5.

Sieve Opening (mm)	Cumulative Percent Passing
4.75	100
2.0	80
0.425	60
0.25	30
0.15	10
0.075	5

So, $D_{60} = 0.425$ mm; $D_{30} = 0.25$ mm; $D_{10} = 0.15$ mm

$$c_u = \frac{D_{60}}{D_{10}} = \frac{0.425}{0.15} = 2.83$$

$$c_c = \frac{D_{30}^2}{D_{60} \times D_{10}} = 0.98$$

Solution 6.

$$\Delta H = \frac{C_c H}{1+e_i}\log\left(\frac{p_i + \Delta p}{p_i}\right) = \frac{(0.27)(3)}{1+1.1}\log\left(\frac{98+42}{98}\right) = 0.0597 \text{ m} = 59.7 \text{ mm}$$

Solution 7.

$$C_c = \frac{e_1 - e_2}{\log\left(\dfrac{p_2}{p_1}\right)} = \frac{0.9 - 0.82}{\log\left(\dfrac{200}{100}\right)} = 0.266$$

Chapter 2
Structural Analysis

Dr. Michael Taylor

Note: The following outline notes are a summary of what can be found in all textbooks upon this subject. Since no one person's choice of summary can hope to match the needs of all students simultaneously, it is recommended that a favorite text be re-read in conjunction with these notes and that personal notes be appended to the text below.

Determinate structural analysis, often termed Statics, deals with structures which do not move. There are only two types of motion - translation and rotation. If the structure (and all parts of it) neither translates nor rotates it is said to be in **static equilibrium** (it is true, of course, that any material under stress will undergo some change of size and/or shape because of those stresses but these movements are considered negligible in the present context).

In statics translation is caused by **forces** and rotations are caused by **moments.** These are vector quantities. To define a vector requires three characteristics - these are usually (a) a line of action (b) a direction and (c) a magnitude. In some contexts moments whose vectors are out of the plane are termed torques or twists. The student is expected to be familiar with simple vectors and their manipulation.

"Actions" is a general term which includes both forces and moments.

Newtons Laws:

The study of statics is based upon two of Newton's three laws of motion. These are (1) a body will remain in its state of rest (relative to some chosen reference point), or of motion **in** a straight line, unless acted upon by a force, and (2) to every action there is an equal and opposite reaction. The third law (a body under the action of a single force will translate with acceleration along the line of the force and in the same direction as the force) is used in the study of Dynamics.

Law number 2 is often restated as "forces (actions) can exist only in equal and opposite pairs". The combination of laws 1 and 2 thus require that the net action on a body at rest be zero.

This, in turn, defines the two vector equations
net force = 0
net moment = 0
This is the central principle of statics.

More generally, determinate structural analysis can be defined as the process of calculating (for a body in static equilibrium)

(a)any one, or all, of the actions acting upon the body
(b)the movement (deflection) of any point (in any direction) within the body
(c) the rotation of any line in the body.

The process of structural analysis has just two steps. **First** an idealized model of the structure (including the forces acting upon it) is agreed upon. In all but the most complex or large structures it is adequate to use two dimensional representations of real structures. When modeling the structure it is also idealize the supports. Such idealizations as frictionless rollers, fully fixed supports and frictionless pins are assumed to be familiar to the reader.

One other assumption is that structures are composed of rigid members. This does not mean that each member is infinitely rigid (i.e. does not change length or shape when loaded) but rather that these deformations are small relative to the original dimensions. Expressed another way, the deformations induced by loading do not change the geometry of the structure significantly. Luckily this assumption is a good one for the vast majority of real engineering structures. The discerning reader may be about to object that many textbook examples (and reality) contain structures with ropes or wires which are not stable if placed in compression. This objection is a good one but is countered by the observation (often implicit) that in all such cases the rope or wire must be in tension and hence cannot "buckle" or otherwise move.

Most engineers drawings (and most examination problems) already include the simplifications described above.

The **second** step of the analysis process is the application of Newtons laws to the model.

The application of these equations of statics (often termed the equations of equilibrium) to a three dimensional body involves just two vector equations (translation = 0 and rotation = 0). However it is frequently more convenient to use the equivalent 6 independent scalar equations (three for translation and three for rotations). For two dimensional models these reduce to 3 (two translations and one rotation). Thus if a two dimensional structural problem involves more than three unknowns it is statically indeterminate, and requires further knowledge for its solution. In what follows the discussion is limited to 2 dimensions and stable static strictures.

Free Body Diagrams

The single most important concept in the solution of statics problems is that of the Free Body Diagram (FBD). This concept is founded on the observation that if a (rigid) structure is in equilibrium then every part of it must also be in equilibrium. This method has proven to minimize the likelihood of errors in solving structural problems.

In this approach an imaginary closed line is drawn through the structure. This line isolates a part of the structure (the part is called the Free Body). The isolated part may be drawn by itself but to maintain consistency with the original structure and loading the engineer must examine the complete length of the closed line and insert all possible actions in their correct form and at their correct locations. Of course some of these actions will be known and some unknown. All known actions must be included with their correct locations, magnitudes and orientations. All unknown actions are given names and their directions may be assumed (the mathematics will provide both the correct magnitudes and the correct directions for each unknown).

It should be obvious that if the FBD is incorrect in **any way** then the solution will be incorrect (a solution may be obtained but, if so, it is the solution to some other problem).

It is recommended that for each FBD used, a clear choice of axes be indicated and that a positive direction also be chosen. These choices are, of course, arbitrary. It is then recommended that when applying the equations of equilibrium all terms be written on one side of the equation and the result set equal to zero. The consistent use of this technique will reduce errors (the practice, for example, of setting "all up forces equal to all down forces" which may seem attractive at first and will certainly yield correct answers, is fraught with peril).

Facility in the analysis of structures is best obtained by solving multiple problems of as wide a variety as can be found. An excellent learning tool is to outline the steps of solution without necessarily performing the mathematics. This permits a more rapid aquisition of the skills needed to understand how new problems are solved.

Note that in the following examples units are not specified. The reader may select any units desired (e.g. metric versus "English" and small versus large).

In the following two problems the entire structure is used as the FBD.

Example 1: Find the reactions necessary to keep the structure shown in equilibrium.
(FBD of entire Structure)

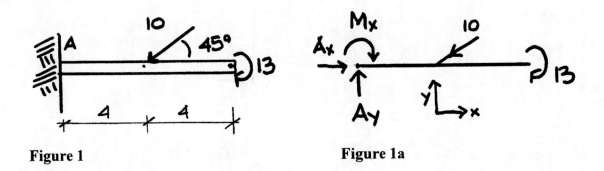

Figure 1   **Figure 1a**

Assume the axes, names and directions shown in the figures.
Step 1. Sum forces in x direction

$$A_x + (-).707 * 10 = 0$$
$$A_x = + 7.07 \text{ (assumed direction correct)}$$

Step 2. Sum forces in y direction

$$A_y + (-).707 * 10 = 0$$
$$A_y = +7.07 \text{ (assumed direction correct)}$$

Step 3. Sum moments about an axis through A (assume clockwise positive)

$$(+) M_x + (+) 10 * 4 * .707 + (+) 13 = 0$$

$$M_x = - 41.3 \text{ (assumed direction was incorrect)}$$

This completes the solution requested.

Example 2: Find the reactions necessary to keep the structure shown in equilibrium.

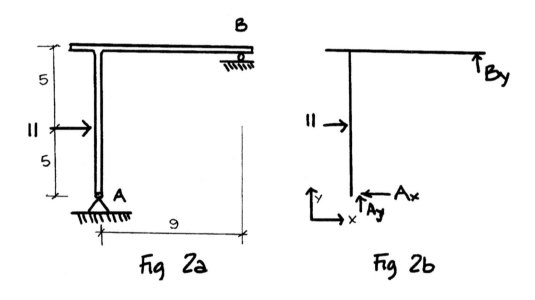

Fig 2a   Fig 2b

(FBD entire Structure)

Assume the names, axes and directions shown.
Step 1. Summing forces in the positive x direction gives

$$+11 + (-)(A_x) = 0$$

Hence $(A_x) = +11$ (i.e. assumed direction is correct)
Step 2. Taking moments about an axis through A (with clockwise positive) gives

$$(-)B_y * 9 + (+)11 * 5 = 0$$

Hence $B_y = + 55/9 = + 6.1$
Step 3. Summing forces in the v direction

$$A_y + 6.1 = 0$$
$$A_y = -6.1 \text{ (i.e. assumed direction was incorrect)}$$

This completes the solution requested.

Trusses and Frames

It should already be obvious to the reader that there is no single best way to solve structural analysis problems. It is true that a judicious choice of FBD and the selection of the order in which equations are solved can shorten the calculation process but all solution approaches will produce the correct answers if applied without error. However it is convenient (but nothing more) to classify certain groups of structures because they share characteristics and thus can be handled by a more "systematic" solution process. Two such groups are trusses and frames.

Trusses

Trusses must meet three criteria:

All members (often termed bars) of the structure must be connected at only two points (these are called joints).
All joints are frictionless pins.
Loads are applied only at the joints.

The FBD of a bar of a truss can thus have only two forces acting upon it - one at each frictionless pin. Note that the bars need not be straight (although they almost always are). Application of the equations of statics shows that the line of action of both these forces must be directed along the line joining the two joints and that they must be equal in magnitude but oppositely directed. Hence a member of a truss is often described as a "two force member". Only two possibilities then exist - the two forces are either directed toward each other (in which case the bar is in compression) or they are directed away from each other (the bar is in tension).

This feature of trusses simplifies their analysis because if a FBD is drawn which cuts a member then it is known that only one action can exist at the cut - it must be a force directed along the line which connects the joints (the sense and magnitude to be found as part of the solution).

There are two "classes" of analysis for trusses.

Method of Joints.

In this approach all FBDs are joints (with the exception that the entire body FBD is often used to find the reactions).

Method of Sections.

In this method the FBDs can include portions of the structure which are larger than a single joint.

It should be noted that most designers use a combination of these methods when solving structures.

Example 3. Find the bar force in member CB using the method of joints.

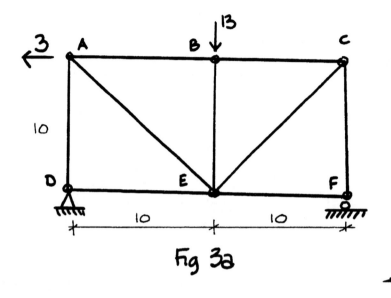

Fig 3a

Figures for Example 3.

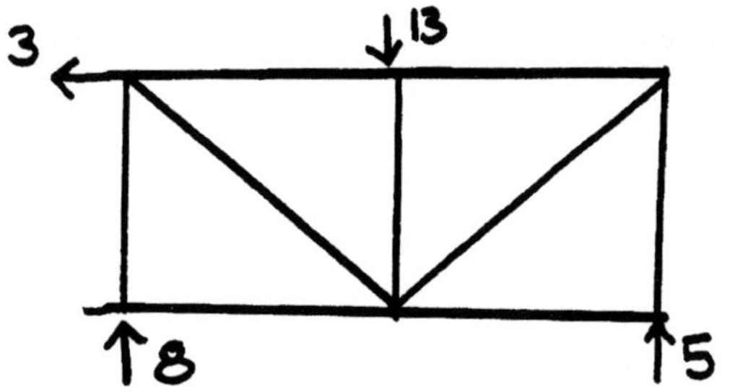

Fig 3b

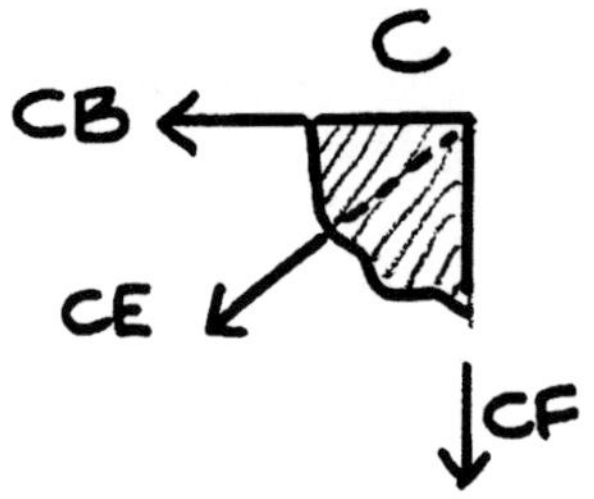

Fig 3c

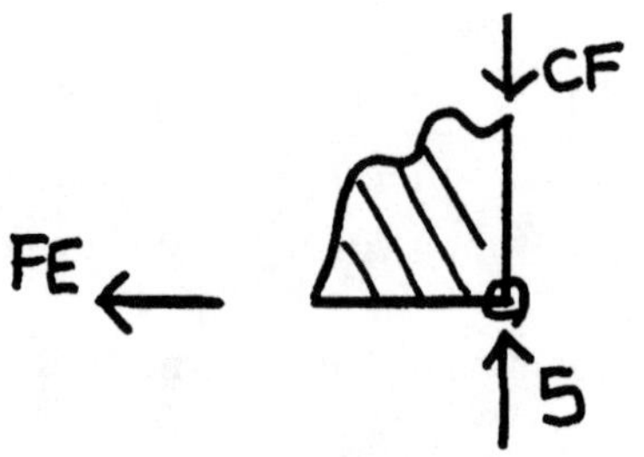

Fig 3d

Use a FBD of the entire structure to obtain reactions (Fig 3a). Assume the axes, names and directions shown.

Solving by methods shown above the reader should obtain

D_y = 8.0 up

F_y = 5.0 up

D_x = 3 to right

This completes reactions as shown in Fig 3b.

Note that if we wish to attempt to find CB using a FBD of joint C there will be 3 unknowns which is not solvable (see Fig 3c). We must find one of the other bar forces, CE or CF in order to obtain a solution for CB. One possibility is to solve joint F first. This will give us CF and then permit a solution at joint C.

With Fig 3d and the assumed directions, joint F is solved to give CF = 5 (with bar in compression).

Now moving to joint C (back to Fig 3c but with CF known in magnitude and direction hence the arrow for CF in Fig 3c must be reversed) and taking moments about E with clockwise assumed positive,

CB = 5 and CB is in compression.

Example 4. Find the bar force in member DH using the method of sections.

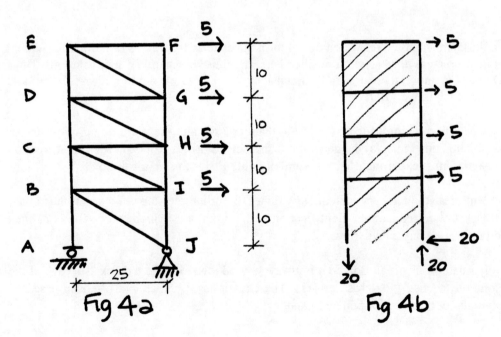

Fig 4a   Fig 4b

(FBD entire Structure)

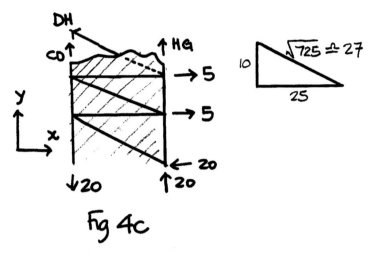

Fig 4c

(FBD with horizontal cut through DC and GH.)

Solve to get the reactions shown in Fig 4b.
Step 1. Note that a horizontal cut can be used to give the FBD shown in Fig 4c.
Sum forces in x direction

$$+5 + 5 - 20 - DH * (25/27) = 0$$
$$DH = +10.8 \text{ (assumed direction correct, bar DH in tension)}$$

Frames.

Frames are multi-member structures which do not meet the criteria for trusses (although individual members within a structure may be two force members, and thus help simplify their solution). Thus, when a FBD cuts a member, there are (in 2-dimensions) three possible actions across the cut and these must be included in the FBD.

Frames are clearly more complicated of solution than trusses and there are no simple rules or guidelines for solution other than (a) the general rule that several FBDs will be needed for a solution and (b) the designer must experiment with several possible FBDs to find a solution.

It is an excellent idea (though very infrequently done) to outline the steps of a solution (i.e. the sequence of FBDs to be used and the identification of the unknowns which can be found at each step) prior to performing any calculations.

When relatively small problems are solved it is always a good idea to make a quick "survey" of the completed solution to check for gross errors. The equations of statics can often be assessed approximately without the need for pen and paper.

Example 5

Determine the force in member AC in figure 5a.
Note that this structure is externally indeterminate - i.e. the reactions cannot be found by the equations of statics alone. However we note that bar AC is a two force member and so we can take BCD as a FBD and solve for the force AC. (note that we cannot solve the second FBD completely but we can get the bar force asked for).

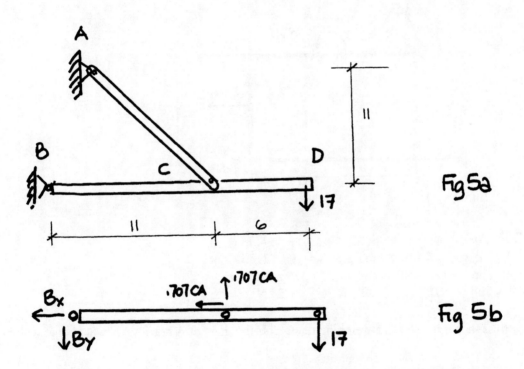

From Fig 5b. and taking moments about B (note that neither of the translation equations will produce a numerical result).

With clockwise positive

(+) 17 * 17 + (-).707 * CA * 11 = 0

CA = 37.2 (assumed direction correct, CA in tension)

Problems:

The reader is encouraged to attempt these problems alone prior to reviewing the sample solutions, and to gain insight by comparing solutions where these differ.

In all cases the self weight of the structure may be ignored unless instructions indicate otherwise. ·

Indicate the best answer from those offered (round off errors may lead to small discrepancies)

2.1 For the structure shown the reactions are (in kN)

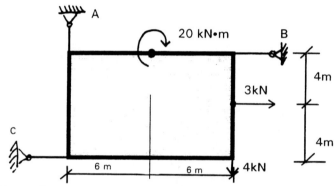

Figure 2-1

(A) A = 4 kN down   B = 10 kN to left   C = 11.5 kN to right
(B) A = 3 kN down   B = 10 kN to right C = 10.0 kN to right
(C) A = 4kN up   B = 10 kN to left   C = 7.0 kN to right
(D) A=4kN up   B = 8.5 kN to left C = 11.5 kN to left

2.2 The truss shown is used as a weighing device. If the gage at A reads 9 kN what is the weight of mass P?

(A)   9 kN
(B)   12kN
(C)   4 kN
(D)   8 kNn

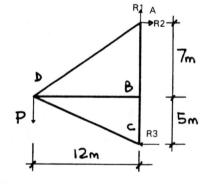

Figure 2-2

2.3 The force in member CB is

(A)4.65 kN tensile
(B)4.65 kN compressive
(C)47.5 kN tensile
(D)8.00 kN compressive

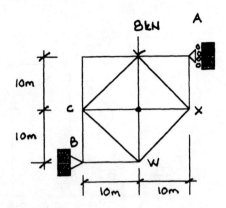

Figure 2-3

2.4 The force in member DE is

(A)   11.2 kN tensile
(B)   11.3 kN compressive
(C)   zero
(D)   none of the above

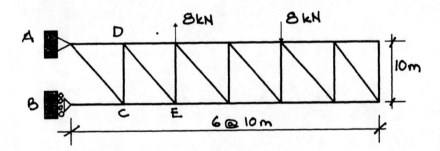

Figure 2-4

2.5 The force in members FB and GH are

(A)   FB = 2.5 kN tensile GH = 11.66 kN tensile
(B)   FB = zero kN tensile GH = 11.66 kN tensile
(C)   FB = zero kN tensile GH = 11.66 kN compressive
(D)   FB = zero kN tensile GH = 10.00 kN tensile

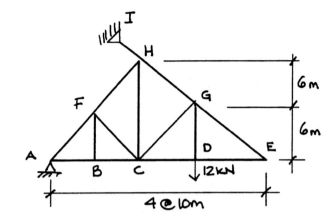

Figure 2-5

Solutions:

2.1

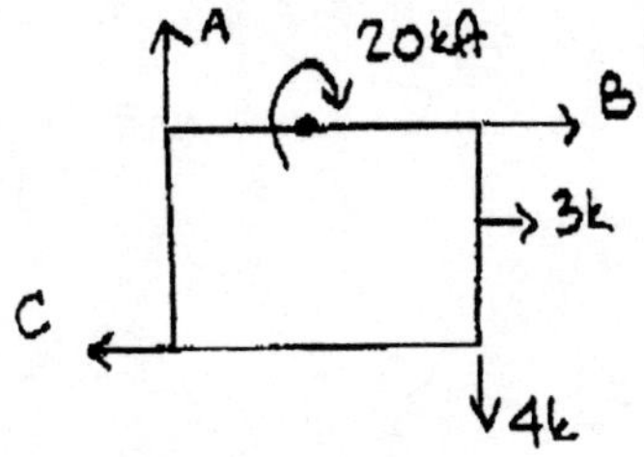

FBD whole struct

(a) $\sum V = 0 \quad \uparrow +$

$\qquad + A - 4 = 0 \qquad \underline{A = 4 \text{ up}}$

(b) $\sum M_c = 0 \quad \circlearrowright$

$\qquad + 20 + B(8) + 3(4) + 4(12) = 0$

$\qquad\qquad B = -10 \quad \text{ie } \underline{B = 10 \leftarrow}$

(c)

$\qquad \sum H = 0 \quad \rightarrow +$

$\qquad -C \; (\rightarrow)10 + 3 = 0 \qquad C = -7, \text{ ie } \underline{C = 7. \rightarrow}$

2.2

$\sum M_c \quad \circlearrowright$

$\qquad + 9 \times 12 - P(12) = 0 \qquad \underline{P = 9k}$

2.3

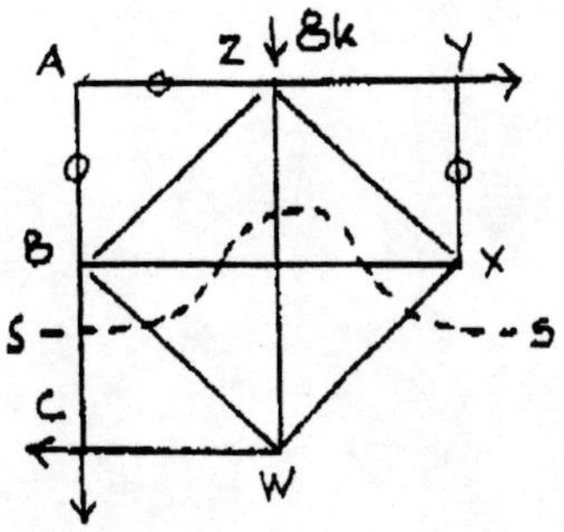

Note $XY = 0 \therefore = ZA = AB$

(a) FBD all

$\qquad \sum M_c \quad \circlearrowright$

$\qquad + 8(10) + R_y(20) = 0 \quad \underline{R_y = 4 \leftarrow}$

$\qquad\qquad \text{so } YZ = 4 \text{ comp.}$

$\sum H \text{ gives } \quad C_H = 4 \rightarrow$

$\sum V \qquad\qquad C_V = 8 \uparrow$

$\qquad \text{so } CB = 8c$

$\qquad\qquad CW = 4c$

2.4

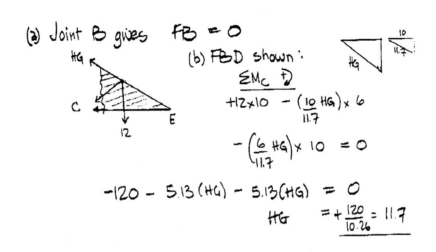

FBD shown

$\Sigma V.$

$+8 - 8 + DE(0.7) = 0 \quad \underline{DE = 0}$

2.5

(a) Joint B gives $FB = 0$

(b) FBD shown :

$\Sigma M_C \; \circlearrowleft$

$+12 \times 10 - \left(\dfrac{10}{11.7} HG\right) \times 6$

$- \left(\dfrac{6}{11.7} HG\right) \times 10 = 0$

$-120 - 5.13(HG) - 5.13(HG) = 0$

$HG = +\dfrac{120}{10.26} = 11.7$

Chapter 3
Hydraulics and Hydro Systems
Dr. Bruce E. Larock

This chapter will review the two equations that are likely to be found in the FE examination. A more thorough presentation may be found in the Civil Engineering License Review or in selected references at the end of this chapter.

Manning Equation

The Manning equation for the average velocity V in a steady open channel flow is

$$V = \frac{K}{n} R^{2/3} S^{1/2}$$

With SI units K = 1.0; with English units K = 1.49. The hydraulic radius is R= A/P with A = flow cross-sectional area and P = wetted perimeter, i.e. the length of the interface along the fluid/solid boundary containing it. The slope S of the energy line is equal to the amount of head lost in a channel section of length L, or $S = h_L/L$. A short table of values for n, the Manning roughness factor, follows:

Channel Surface	Roughness Value n
Concrete, finished	0.012
Concrete, gunite	0.019
Clay, vitrified sewer	0.014
Rubble masonry	0.025
Concrete, mortar	0.013
Concrete, troweled	0.013
Gravel, clean	0.025

Although the Manning equation is intended primarily for use in open channels, it is sometimes also used for pressurized flow in pipes.

Example 3-1
A rectangular open channel lined with rubble masonry is 5 meter wide and laid on a slope of 0.0004. If the depth of uniform flow is 3 meter, compute the discharge in m³/sec.
Solution
From the table the Manning roughness is approximately n = 0.025. The area and wetted perimeter are
A = (5)(3) = 15 m², P = 5 + 2(3) = 11 m, and R = A/P
The Manning equation now gives

$$Q = AV = A\left(\frac{1}{n}\right)R^{2/3} S^{1/2} = (15)\left(\frac{1}{0.025}\right)(15/11)^{2/3}(0.0004)^{1/2}$$

Q = 14.76 m³/sec

Example 3-2

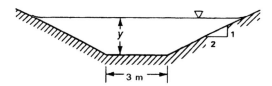

A gunite concrete trapezoidal channel with 1:2 side slopes, which is shown in Fig. 6-12, conveys 60 m^3/sec on a slope $S_o = 0.0005$. Compute the depth of uniform flow.

Solution

In SI units the Manning equation is

$$Q = (1/n) \, AR^{2/3} \, S_o^{1/2}$$

From the table the appropriate roughness coefficient is $n = 0.019$. Inserting the given information leads to

$$AR^{2/3} = 5.10 \qquad\qquad \text{(a)}$$

in which

$$A = 3y + 2y^2$$
$$P = 3 + 2\sqrt{5}\, y \quad \text{and} \quad R = A/P$$

Equation (a) must now must be solved by successive trial. It is convenient to use a table in doing so:

Trial	y(m)	A(m²)	P(m)	R	R²/³	AR²/³ = 5.10?
1	1.5	9.00	9.71	0.93	0.95	8.55
2	1.0	5.00	7.47	0.67	0.77	3.85
3	1.15	6.10	8.14	0.75	0.83	5.06
4	1.16	6.17	8.19	0.75	0.83	5.12

The normal depth, that is, the depth of uniform flow, is $y = 1.16$ m.

Hazen-Williams Equation

Many empirical formulas for pipe friction have been developed over the past century. These formulas are usually based on tests involving the flow of water under fully turbulent conditions and are not normally reliable for use with other fluids. One relatively widely-used such formula is the Hazen-Williams equation, which is

$$V = 0.849 \, CR^{0.63} \, S^{0.54}$$

for SI units. For English units, replace 0.849 with 1.318. The Hazen-Williams coefficient C ranges from approximately 140 for very smooth and straight pipes, to 120 for smooth masonry, to 100 or less for old cast iron pipe. The other factors in the equation are defined as they were for the Manning equation.

Example 3-3

If 0.01 m^3/sec of water flows through a new 100 mm clean cast iron pipe (C= 130), determine the head loss in 1000 m of this pipe.

The discharge Q = *VA* with

$$A = \frac{\pi}{4} (0.1)^2 = 0.00785 m^2$$

Hence V = 0.01/0.00785 = 1.273 m/sec
In this case the hydraulic radius is

$$R = \frac{\pi \dfrac{D^2}{4}}{\pi D} = \frac{D}{4} = \frac{0.1}{4} = 0.025 \text{ m}$$

The Hazen-Williams equation then yields
V = 1.273 = 0.849 (130) (0.025)$^{0.63}$ S$^{0.54}$
S= 0.0191 = h$_L$/L = h$_L$/1000 and h$_L$ = 19.1m

References

Chow, V.T. *Open Channel Hydraulics*. McGraw-Hill, New York,1959.

Henderson, F. M. *Open Channel Flow*. Macmillan, New York,1966.

Street, R. L., Watters, G. Z. and J. K. Vennard, *Elementary Fluid Mechanics, 7th Ed.*, Wiley, New York, 1996.

White, F.M. *Fluid Mechanics*, 3rd Ed. McGraw-Hill, New York, 1994.

3-4

Chapter 4
Structural Steel and Reinforced Concrete Design
Dr. Alan Williams

Elastic Design of Steel Beams

The allowable stress on flexural members depends on the shape of the section and the bracing used to prevent lateral instability[1]. Sections are classified as compact, noncompact and slender in accordance with the criteria given in AISC Table B5.1. Most rolled W shapes qualify as compact sections. The exceptions are indicated in the AISC Tables of Properties by denoting the value of the yield stress F_y' and F_y''' at which a particular shape becomes noncompact. The allowable bending stress for compact symmetrical shapes is given by AISC Equation (F1-1) as

$$F_b = 0.66\, F_y$$

when the maximum unbraced length of the compression flange does not exceed the smaller of

$$L_c = 76 b_f / \sqrt{F_y} \quad \text{or} \quad 20{,}000 / (dF_y / A_f)$$

where b_f = flange width, and A_f = compression flange area.
When the unbraced length of a compact shape exceeds L_c, but is less than the greater of

$$L_u = r_T \sqrt{102{,}000 C_b / F_y} \quad \text{or} \quad 20{,}000\, C_b / (dF_y / A_f)$$

the allowable bending stress is given by AISC Section F1.3 as
$$F_b = 0.60 F_y$$
where r_T = radius of gyration of the compression flange plus one-third of the web area
C_b = bending coefficient defined in Section F1.3 and Table 6
   = $1.75 + 1.05 M_1 / M_2 + 0.3(M_1 / M_2)^2$, but not more than 2.3.

The value of C_b conservatively may be taken as unity.
Examples of the derivation of C_b are illustrated in Fig. 4-1. Values of L_c and L_u are tabulated in the AISC Beam Tables, assuming a value for C_b of unity.
   In general, the allowable bending stress for noncompact symmetrical shapes is given by AISC Equation (F1-5) as
$$F_b = 0.60\, F_y$$

when the maximum unbraced length of the compression flange does not exceed L_c.
   The AISC Selection Tables and Beam Tables tabulate the allowable beam resisting moments and uniformly distributed loads for W, M, S, C and MC shapes that are braced at the appropriate values of L_c or L_u. Adequate bracing is assumed to be provided by a restraint with a capacity of one percent of the force in the compression flange in accordance with AISC Section G4.

1. Metric steel tables are not available and hence it is not possible to metricate the steel section.

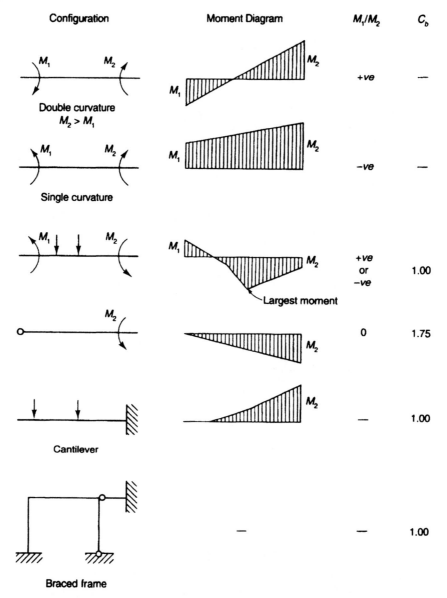

Figure 4-1. Derivation of C_b

When the unbraced length of the compression flange exceeds L_u, the allowable bending stress for both compact and noncompact shapes is given by the larger value from AISC Equations (F1-6), (F1-7) and (F1-8) but may not exceed $0.60\,F_y$. When the unbraced length is less than

$$l = r_T \sqrt{510,000 C_b / F_y} \;\; = L_1$$

the applicable equation is (F1-6) and the allowable stress is

$$F_b = F_y \left(0.667 - F_y l^2 / 1,530,000 r_T^2 C_b\right).$$

When the unbraced length equals or exceeds L_l, the applicable AISC equation is (Fl-7) and the allowable stress is

$$F_b = 170,000C_b r_T^2 / l^2$$

Equation (Fl-8) is independent of l/r_T and gives an allowable stress of

$$F_b = 12,000C_b A_f / ld.$$

These expressions may be solved by calculator, or the allowable beam resisting moment may be obtained from AISC Moment Charts, which are based on a value of unity for C_b and are conservative for larger values of C_b.

Compact sections and solid rectangular sections bent about their weak axes, and solid round or square bars have an allowable stress given by AISC Equation (F2- 1) of

$$F_b = 0.75F_y.$$

Noncompact sections bent about their weak axes have an allowable stress given by AISC Equation (F2-2) of

$$F_b = 0.60F_y.$$

Example 4-1

The W18 x 60 grade A36 beam, shown in Fig. 4-2 is laterally supported throughout its length. Determine if the beam is adequate to support the applied loads indicated. The relevant properties of the beam are $S_x = 108$ inches^3, $F_y = 36$ kips per square inch, and allowable bending stress $F_b = 0.66 F_y$.

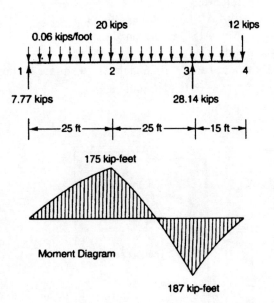

Figure 4-2

Solution

The bending moments acting on the beam from the applied loads and beam self-weight are shown in Fig. 4-2.

M_x =maximum moment = 187 kip-feet,

f_b = maximum bending stress = M_x /S_x = 187 x 12/108 = 20.78 kips per square inch

The allowable stress is

$\qquad F_b = 0.66\ Fy = 0.66$ x 36 = 23.76 kips per square inch.

Hence, the W18 x 60 is adequate.

The allowable shear stress, based on the overall beam depth, is given by AISC Equation (F4-1)

as $\qquad\qquad\qquad\qquad\qquad F_v = 0.40\ F_y$

provided

$$h / t_w \le 3\,80\ /\ \sqrt{F_y}$$

and the actual shear stress is determined by

$$f_v = V / dt_w\ .$$

where h = clear distance between the flanges, t_w = web thickness, d = overall depth of beam, and V = applied shear force.

When the end of the beam is coped, failure occurs by block shear, or web tear-out, which is a combination of shear along a vertical plane and tension along a horizontal plane. The resistance to block shear is given by

$$V_B = A_v F_v + A_t F_t$$

where A_v = net shear area, A_t = net tension area,

F_v = allowable shear stress = 0. 30 F_u from AISC Equation (J4- 1), and

F_t = allowable tensile stress = 0.50 F_u from AISC Equation (J4-2).

From Fig. 4-3 where $d_h = d_b + 0.0625$ inches

$A_v = t_w(\ l_v + 2s - 2.5d_h) = 0.31\ [1.5 + 6 - 2.5(0.75 + 0.0625)] = 1.70$ square inches

$A_t = t_w\ (l_h - 0.5d_h) = 0.31\ (1.5 - 0.5$ x $0.8125) = 0.34$ square inches.

For grade A36 steel, $F_u = 58$ kips per square inch.

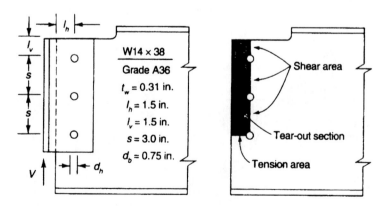

Figure 4-3. Block shear in a coped beam

The resistance to block shear is then

$$V_B = A_v F_v + A_t F_t = 1.7 \times 0.30 \times 58 + 0.34 \times 0.50 \times 58 = 39.44 \text{ kips.}$$

Compression Members

The allowable stress in an axially loaded compression member is dependent on the slenderness ratio which is defined in AISC Section E2 as Kl / r, where r = the governing radius of gyration Kl = effective length of the member, K = effective-length factor, and l = unbraced length of the member.

End Restraints	Ideal K	Practical K	Shape
Fixed at both ends	0.5	0.65	
Fixed at one end, pinned at the other end	0.7	0.8	
Pinned at both ends	1.0	1.0	
Fixed at one end with the other end fixed in direction but not held in position	1.0	1.2	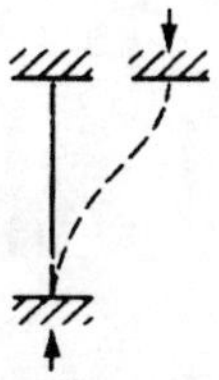
Pinned at one end with the other end fixed in direction but not held in position	2.0	2.0	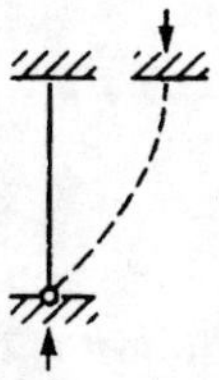
Fixed at one end with the other end free	2.0	2.1	

Figure 4-4 Effective-length factors

The value of the effective-length factor depends on the restraint conditions at each end of the column. AISC Table C-C2.1 specifies effective-length factors for well defined, standard conditions of restraint, and these are illustrated in Fig. 4-4. Values are indicated for ideal and practical end conditions, allowing for the fact that full fixity may not be realized. These values may only be used in simple cases when the tabulated end conditions are approached in practice.

For compression members in a plane truss, AISC Section C-C2 specifies an effective-length factor of 1.0. For load bearing web stiffeners on a girder, AISC Section K1.6 specifies an effective-length factor of 0.75. For columns in a rigid frame which is adequately braced, AISC Section C2.1 specifies a conservative value for the effective-length factor of 1.0.

The failure of a short, stocky column occurs at the squash load when the strut yields in direct compression. The allowable stress in the strut is obtained from AISC Equation (E2-1) as

$$F_a = 0.6\,F_y$$

As the slenderness ratio of the column is increased, the failure load reduces. The Euler elastic critical load is assumed to govern when the column stress equals half the yield stress. The critical slenderness ratio, corresponding to this limit, is given by AISC Equation (C-E2-1) as

$$C_c = \sqrt{2\pi^2\,E\Big/F_y}$$

When the slenderness ratio exceeds this value, the allowable stress is
$$F_a = 12\pi^2 E\big/23(\,Kl\,/\,r\,)^2$$

When the slenderness ration does not exceed this value, the allowable stress may be obtained from the expression
$$F_a = [1-(Kl/r)^2/2C_c^2]F_y\,/[5/3 +3(Kl/r)/8C_c -(Kl/r)^3/8C_c^3]$$

In accordance with AISC Section B7, the slenderness ratio should preferably not exceed 200.

Example 4-2

The W14 x 120 Grade A36 column, shown in Fig. 4-5, is fixed at the base and unbraced about the x-axis. About the y-axis, the column is held in position, at the top and at mid height, but it is not fixed in direction. Determine if the column is adequate to support an axial load of 500 kips. The relevant properties of the W 14x120 are $A = 35.3$ inches^2, $r_x = 6.24$ inches, $r_y = 3.74$ inches, $E_s = 29000$ ksi, $F_y = 36$ ksi, $K_y = 0.8$, and $K_x = 2.1$.

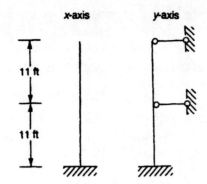

Figure 4-5. Column in example 4-2

Solution

From Fig. 4-4, the slenderness ratio about the y-axis is given by
$Kl/r_y = 0.8 \times 11 \times 12/3.74 = 28.2$.
The slenderness ratio about the x-axis is given by
$Kl/r_x = 2.1 \times 22 \times 12/6.24 = 88.8$ which governs.

$$C_c = \sqrt{2\pi^2 \times 29{,}000/36} = 126.1 > kl/r_x$$

The allowable axial stress is
$F_a = [1-88.8^2/(2 \times 126.1^2)]36/[5/3 + 3 \times 88.8/(8 \times 126.1) - 88.8^3/(8 \times 126.1^3)] = 14.34$ ksi.
The axial stress from the imposed loading is
$f_a = P/A = 500/35.3 = 14.16$ kips per square inch. $< F_a$, The column is adequate.

In determining the capacity of a connection in direct tension, as shown in Fig. 4-6, allowance must be made for the effective areas of the members and their method of attachment. To prevent excessive elongation of the members, which may lead to instability of the whole structure, AISC Section D1 limits the maximum tensile force on the connection to

$$P_t = 0.6F_y A_g$$

where A_g = gross area of the member = bt.

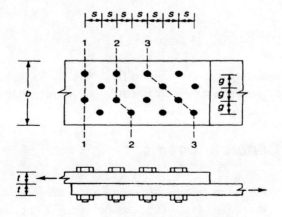

Figure 4-6. Net area of tension member

To prevent fracture of the member at the section of weakest effective net area, AISC Section D1 limits the maximum tensile force on the connection to

$$P_t = 0.5 F_u A_e$$

where A_e = effective net area. The effective net-area is defined in AISC Section B2 as:

$$A_e = t(b - 2d_h), \text{ for Section 1-1}$$

where d_h = specified diameter of hole =$(d_b + 1/8\text{-inch})$, where d_b = bolt diameter:

$$A_e = t(b - 3d_h + s^2/4g) \text{ for Section 2-2 .}$$

where s = longitudinal pitch, and g = transverse gage;

$$A_e = t(b - 4d_h + 3s^2/4g) \text{ for Section 3-3 .}$$

To account for the effects of eccentricity and shear lag in rolled structural shapes connected through only part of their cross-sectional elements, the effective net area is given by AISC Equation (B3-1) as

$$A_e = UA_n$$

where A_n = net area of the member,
U= 0.90 for I-Sections with $b_f \geq 2d/3$ and with not less than three bolts in the direction of stress
U= 0.85 for all other shapes with not less than three bolts in the direction of stress, and
U= 0.75 for all shapes with only two bolts in the direction of stress.
In addition, AISC Section B3 specifies that $A_e \leq 0.85 A_g$.

Strength Design Principles for Reinforced Concrete Members

The basic requirement of designing for strength is to ensure that the design strength of a member is not less than the required ultimate strength. The latter consists of the service-level loads multiplied by appropriate load factors, and this is defined in ACI Equations[2] (9-1), (9-2) and (9-3) as

$$U = 1.4D + 1.7L$$
$$U = 0.75(1.4D + 1.7L + 1.7W)$$
$$U = 0.9D + 1.3W$$

where D = dead load, L = live load, and W = wind load.
The design strength of a member consists of the theoretical ultimate strength of the member-the nominal strength-multiplied by the appropriate strength reduction factor, ϕ.

Thus

$$\phi(\text{nominal strength}) \geq U$$

ACI Section 9.3 defines the reduction factor as ϕ = 0.90 for flexure, ϕ =0.85 for shear and torsion, ϕ = 0.75 for compression members with spiral reinforcement, ϕ = 0.70 for compression members with lateral ties, and ϕ = 0.70 for bearing on concrete.

Flexure of Reinforced Concrete Beams

The nominal strength of a rectangular beam, with tension reinforcement only, is derived from the assumed ultimate conditions shown in Fig. 4-7. ACI Section 10.2.7.1 specifies an equivalent rectangular stress block in the concrete of $0.85 f_c'$, with a depth of

$$a = A_s f_y / 0.85 f_c' b = \beta_1 c$$

where c = depth to neutral axis, and β_1 = compression zone factor given in ACI Section 10.2.7.3.

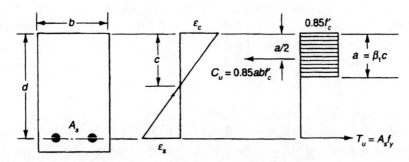

Figure 4-7 Member with tension reinforcement only

From Fig. 4-7, the nominal strength of the member is derived as

$$M_n = A_s f_y d(1 - 0.59 \rho f_y / f_c')$$

where $\rho = A_s /bd$ = reinforcement ratio, $\phi M_n = 0.9 M_n$ = design strength, and $\phi M_n \geq M_u$ = applied factored moment. This expression may also be rearranged to give the reinforcement ratio required to provide a given factored moment, M_u, as

$$\rho = 0.85 f_c' \left[1 - \sqrt{1 - K / 0.383 f_c'} \right] / f_y$$

where $K = M_u / bd^2$.

These expressions may be readily applied using standard calculator programs[5] and tables[3,4]. For a balanced strain condition, the maximum strain in the concrete-and in the tension reinforcement- must simultaneously reach the values specified in ACI Section 10.3.2 as

$$\varepsilon_c = 0.003 = \text{concrete strain}$$
$$\varepsilon_s = f_y / E_s = \text{steel strain}$$

The balanced reinforcement ratio is given as $\rho_b = 0.85 \times 600 \, \beta_1 f_c' / f_y (600 + f_y)$.

In accordance with ACI Section 10.3.3, the maximum allowable reinforcement ratio to ensure a ductile flexural failure with adequate warning of impending failure is

$$\rho_{max} = 0.75 \, \rho_b$$

The maximum allowable reinforcement area is $A_{max} = bd\rho_{max}$
The maximum allowable design strength of a singly-reinforced member is

$$M_{max} = \phi A_{max} f_y d(1 - 0.59 \, \rho_{max} f_y / f_c')$$

In accordance with ACI Sections 10.5.1. and 10.5.2, the minimum allowable reinforcement ratio is given by

$$\rho_{min} = 1.4 / f_y$$

with the exception that the minimum reinforcement provided need not exceed one-third more than required by analysis.

Example 4-3

A reinforced concrete beam, with an overall depth of 400 mm, an effective depth of 350 mm, and a width of 300 mm, is reinforced with Grade 400M bars and has a concrete cylinder strength of 21 MPa. Determine the area of tension reinforcement required for the beam to support a superimposed live load of 15 kN/m over an effective span of 6 m.

Solution

The weight of the beam is $w_D = 0.4 \times 0.3 \times 23.5 = 2.82$ kN/m

The dead load moment is given by $M_D = w_D \ell^2/8 = 2.82 \times 6^2/8 = 12.7$ kNm

The live load moment is given by $M_L = w_L \ell^2/8 = 15 \times 6^2/8 = 67.5$ kNm

The factored moment is $M_u = 1.4 M_D + 1.7 M_L = 1.4 \times 12.7 + 1.7 \times 67.5 = 132.5$ kNm

The moment factor is $K = M_u/bd^2 = 132.5 \times 10^6/(300 \times 350^2) = 3.605$ Mpa

The reinforcement ratio required to provide a given factored moment M_u is

$\rho = 0.85 \, \text{fc}' \, [1-(1 - K/0.383 \, \text{fc}')^{0.5}] / f_y = 0.85 \times 21 \{1 - [1 - 3.605/(0.383 \times 21)]^{0.5}\} /400 = 0.011$ The

minimum allowable ratio is $\rho_{min} = 1.4/f_y = 1.4/400 = 0.0035 < \rho$ ... satisfactory

The maximum allowable reinforcement ratio is $\rho_b = 0.75 \times 0.85 \times 600\beta_1 \, \text{fc}' / f_y(600 + f_y)$

$= 0.75 \times 0.85 \times 600 \times 0.85 \times 21/400(600 + 400) = 0.017 > \rho$ ... satisfactory

Hence the section size is adequate.

The reinforcement area required is $A_s = \rho bd = 0.011 \times 300 \times 350 = 1155$ mm²

When the applied factored moment exceeds the maximum design strength of a singly-reinforced member that has the maximum allowable reinforcement ratio, compression reinforcement, and additional tensile reinforcement must be provided, as shown in Fig. 4-8. The difference between the applied factored moment and the maximum design moment strength of a singly reinforced section is $M_r = M_u - M_{max}$ = residual moment.

The additional area of tensile reinforcement required is

$$A_T = M_r / \phi f_y (d - d') = A_s' f_s' / f_y$$

The depth of the stress block is $a = f_y A_{max} / 0.85 f_c' b$

The depth of the neutral axis is $c = a / \beta_1$

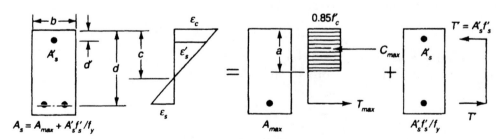

Figure 4-8. Member with compression reinforcement

The stress in the compression reinforcement is given by[5]
$$f_s' = 600(1 - d'/c) \leq f_y$$
The required area of compression reinforcement is given by
$$A_s' = M_r / \phi f_s'(d - d')$$
The total required area of tension reinforcement is
$$A_s = A_{max} + A_s' f_s' / f_y$$
The maximum allowable reinforcement ratio is given by Section 10.3.3 as
$$\rho'_{max} = 0.75 \, \rho_b + A f_s' / b d f_y$$

In order to analyze a given member with compression reinforcement, an initial estimate of the neutral axis depth is required. The total compressive force in the concrete and compression reinforcement is then compared with the tensile force in the tension reinforcement. The initial estimate of the neutral axis depth is then adjusted until these two values are equal.

The conditions at ultimate load in a flanged member, when the depth of the equivalent rectangular stress block exceeds the flange thickness, are shown in Fig. 4-9. The area of reinforcement required to balance the compressive force in the flange is given by
$$A_{sf} = C_f / f_y = 0.85 f_c' h_f (b - b_w) / f_y$$
The corresponding design moment strength is
$$M_f = \phi A_{sf} f_y (d - h_f / 2)$$

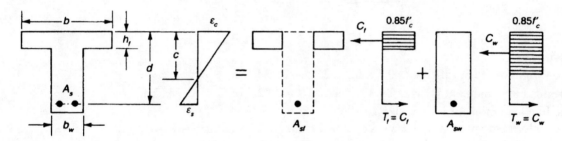

Figure 4-9. Flanged member with tension reinforcement.

The residual moment is $M_r = M_u - M_f$. The required reinforcement ratio to provide the residual moment is

$$\rho_w = 0.85 f_c' \left[1 - \sqrt{1 - 2K_w / 0.9 \times 0.85 f_c'} \right] / f_y$$

where $K_w = M_r / b_w d^2$.

The corresponding reinforcement area is
$$A_{sw} = b_w d \rho_w$$

The total reinforcement area required is
$$A_s = A_{sf} + A_{sw}$$
The total reinforcement ratio is
$$\rho = A_s / bd$$

The maximum allowable reinforcement ratio is given by ACI Section 10.3.3 as

$$\rho'_{max} = 0.75b_w\,(\rho_b + \rho_f)\,/\,b$$
where $\rho_b = 0.85 \times 600\,\beta_1 f_c'\,/\,f_y\,(600 + f_y)$, and $\rho_f = A_{sf}/b_w d$.

Example 4-4

The reinforced concrete beam shown in Fig 4-10 is reinforced with Grade 400M bars at the positions indicated and has a concrete cylinder strength of 21 MPa. The beam carries a superimposed load of 30kN/m run over an effective span of 6m. Determine the areas of tension and compression steel required.

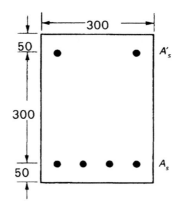

Figure 4-10.

Solution

The weight of the beam is $w_D = 0.4 \times 0.3 \times 23.5 = 2.82$ kN/m

The dead load moment is given by $M_D = w_D\,l^2\,/\,8 = 2.82 \times 6^2\,/\,8 = 12.7$ kNm

The live load moment is given by $M_L = w_L\,l^2\,/\,8 = 30 \times 6^2\,/\,8 = 135.0$ kNm

The factored moment at mid span is obtained from ACI Equation (9-1) as

$$M_u = 1.4M_D + 1.7M_L = 1.4 \times 12.7 + 1.7 \times 135.0 = 247 \text{ kNm}$$

The compression zone factor is given by ACI Section 10.2.7.3 as $\beta_1 = 0.85$.

The maximum allowable reinforcement ratio for a singly reinforced beam is

$\rho_{max} = 0.75 \times 0.85 \times 600 \times \beta_1 f_c'/f_y\,(600 + f_y) = 0.75 \times 0.85 \times 600 \times 0.85 \times 21/400(600 + 400) = 0.017$.

The maximum reinforcement area for a singly reinforced beam is

$$A_{max} = bd\,\rho_{max} = 300 \times 400 \times 0.017 = 2040 \text{ mm}^2.$$

The maximum design moment of a singly reinforced section is

$$M_{max} = \phi A_{max} f_y\,d\,(1 - 0.59\,\rho_{max}\,f_y\,/\,f_c')$$
$$= 0.9 \times 2040 \times 400 \times 135(1 - 0.59 \times 0.017 \times 400/21)/10^6 = 208 \text{ kNm}$$

The residual moment is given by

$$M_r = M_u - M_{max} = 39 \text{ kNm}$$

Depth of stress block $a = f_y A_{max}/0.85f_c'b = 400 \times 2040\,/\,(0.85 \times 21 \times 300) = 152$ mm

Neutral axis depth $c = a\,/\,\beta_1 = 152\,/\,0.85 = 179$ mm

The stress in the compression reinforcement is

$$f_s' = 600(1 - d'/c) = 600 - (1 - 50/179) = 400 \text{ MPa maximum.}$$

The required area of compression reinforcement is

$$A_s' = M_r\,/\,\phi_s f_s'\,(d - d') = 106 \times 39/0.9 \times 400 \times 300 = 361 \text{ mm}^2.$$

The total required area of tension reinforcement is

$$A_s = A_{max} + A_s' f_s'\,/\,f_y = 2040 + 361 = 2401 \text{ mm}^2$$

Deflection Requirements

Allowable deflections are given in ACI Table 9.5(b). For reinforced concrete members not supporting deflection-sensitive construction, the allowable deflection may be deemed satisfied if the minimum thickness requirements of ACI Tables 9.5(a) and 9.5(c) are adopted.

For other conditions, the short-term deflection may be computed from the effective moment of inertia given by ACI Equation (9-7) and indicated in Fig. 4-11 as

$$I_e = (M_{cr} / M_a)^3 I_g + [1- (M_{cr} / M_a)^3] I_{cr}$$

where I_g = moment of inertia of gross concrete section, neglecting reinforcement = $bh^3 / 12$,
I_{cr} = moment of inertia of the cracked transformed section = $b(kd)^3 / 3 + nA_s(d- kd)^2$,
k = neutral axis depth ratio at service load for a singly reinforced section
$\quad = [2\rho n +(\rho n)^2]^{0.5} - \rho n$,
$n = E_s /E_c$ = modular ratio,
$\rho = A_s / bd$ = reinforcement ratio,
M_a = maximum moment in the member,
M_{cr} = cracking moment of the section = $2f_r I_g / h$, and

$f_r = 0.62 \sqrt{f_c'}$ = modulus of rupture of normal weight concrete.

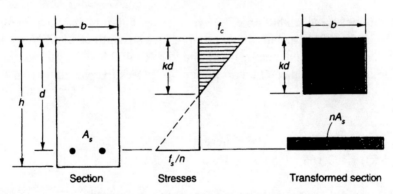

Figure 4-11. Reinforced member at service load

When subjected to long-term loads, creep and shrinkage of the member produces additional deflection, which can be determined by multiplying the short-term deflection by the factor

$$\lambda = \xi / (1 + 50 \rho')$$

where ξ = time-dependent factor for sustained loads given in ACI Section 9.5.2.5, and $\rho' = A_s' / b_w d$ = reinforcement ratio for compression reinforcement.

Shear in Reinforced Concrete Members

The factored shear force acting on a member, in accordance with ACI Equations (11-1) and (11-2), is resisted by the combined design shear strength of the concrete and shear reinforcement. Thus, the factored applied shear is given by

$$V_u = \phi V_c + \phi V_s$$

where $\phi = 0.85$ = strength reduction factor for shear given by ACI Section 9.3.2.3,

$V_c = 0.166\sqrt{f'_c}\ b_w d$ = nominal shear strength of normal weight concrete from ACI (11-3),
$V_s = A_v f_y d /s$ = nominal shear strength of shear reinforcement from ACI Equation (11-17),
A_v = area of shear reinforcement,
s = spacing of shear reinforcement specified in ACI Section 11.5.4 as
  $\leq$ d/2 or 610mm when $V_s \leq .33\sqrt{f'_c}\ b_w d$, and
  $\leq$ d/4 or 305mm when $0.33\sqrt{f'_c}\ b_w d < V_s \leq 0.66\sqrt{f'_c}\ b_w d$.

 The dimensions of the section or the strength of the concrete must be increased, in accordance with ACI Section 11.5.6.8, to ensure that $V_s \not> 0.66\sqrt{f'_c}\ b_w d$

 As specified in ACI Section 11.5.5, a minimum area of shear reinforcement is required when
$V_u > \phi V_c /2 = V_{u(min)}$
and the minimum area of shear reinforcement is given by ACI Equation (11-14) as
$A_{v(min)} = 0.34 b_w s / f_y$

 When the support reaction produces a compressive stress in the member, ACI Section 11.1.3.1 specifies that the critical factored shear force is that which is acting at a distance equal to the effective depth from the face of the support. Flow Chart 4-1 summarizes the shear provisions of the Code, and Fig. 4-12 illustrates the design principles involved.

Example 4-5
A reinforced concrete beam, which has an effective depth of 600 mm and a width of 500 mm, is reinforced with Grade 400M bars and has a compressive strength of 21 MPA. The factored shear force, at four locations on the beam, is (a) 800 kN, (b) 400 kN, (c) 100 kN, and (d) 50 kN. Determine the required spacing, at each location, for No. M10 stirrups with two or four vertical legs as suitable.

Solution
The design shear strength provided by the concrete is
$\phi V_c = 0.166\ \phi b_w d(f_c')^{0.5} = 0.166 \times 0.85 \times 500 \times 600(21)^{0.5}/1000 = 194$ kN

(a) The design shear strength required from the shear reinforcement is
$\phi V_s = V_u - \phi V_c = 800 - 194 = 606$ kN
Since $\phi V_s < 4 \times \phi V_c$ the concrete section is adequate.
Since $\phi V_s > 2 \times \phi V_c$ the maximum stirrup spacing is given by
$s = d/4 = 600/4 = 150$ mm
The area of shear reinforcement required is
$A_v/s = \phi V_s/\phi d f_y = 606 \times 10^6/ (0.85 \times 600 \times 400) = 2970$ mm²/m
The required spacing of stirrups with four No. M10 vertical legs is
$s = 4 \times 100/2.97 = 135$ mm ... < 150 mm ... satisfactory
(b) The design shear strength required from the shear reinforcement is
$\phi Vs = V_u - \phi V_c = 400 - 194 = 206$ kN
Since $\phi V_s < 2 \times \phi V_c$, the maximum stirrup spacing is given by
$s = d/2 = 600/2 = 300$ mm

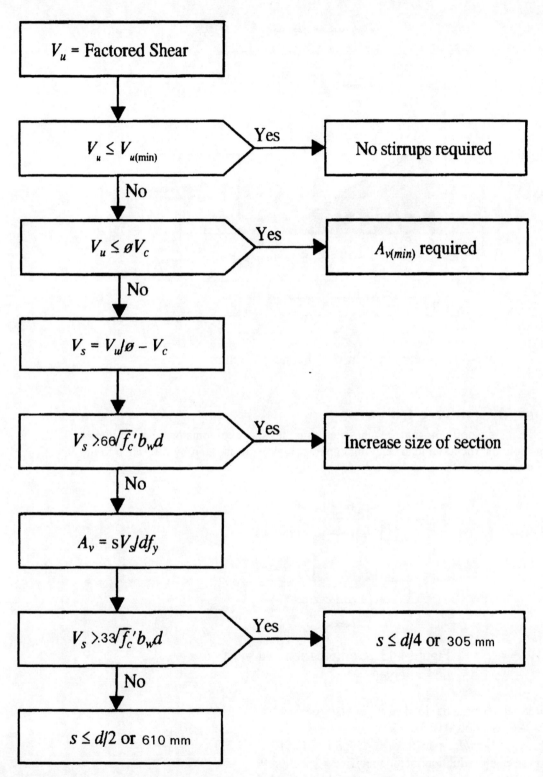

Flow Chart 4-1. Design for shear

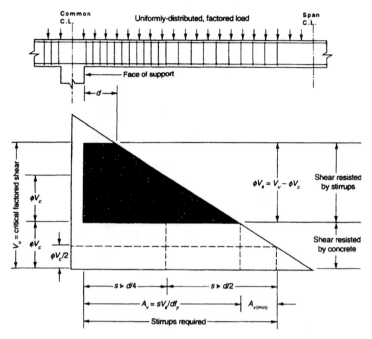

Figure 4-12. Shear in a reinforced concrete beam.

Reinforced Concrete Columns

Reinforced concrete colunms may be classified as either short colunms or long colunms. For a short column, the axial load carrying capacity may be illustrated by reference to the column shown in Figure 4-13.

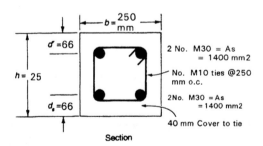

Figure 4-13. Compression in a short tied column.

The theoretical design axial load strength at zero eccentricity is given by ACI Section R10.3.5. as

$$\phi P_o = \phi [0.85 f'_c (A_g - A_{st}) + A_{st} f_y]$$

where ϕ = strength reduction factor specified in Section 9.3

= 0.75 for compression members with spiral reinforcement and

= 0.70 for compression members with lateral ties

f'_c = concrete cylinder strength = 21 MPa

f_y = reinforcement yield strength = 400 MPa
A_g = gross area of the section = 62,500 mm^2
A_{st} = reinforcement area = 2800 mm^2.

Then, $\phi P_o = 0.7[0.85 \times 21(62,500-2800) + 2800 \times 400]/1000 = 1530$ kN

To account for accidental eccentricity, ACI Section 10.3.5 requires a spirally reinforced column to be designed for a mimimum eccentricity of approximately $0.05h$ which gives a maximum design axial load strength at zero eccentricity, in accordance with ACI Equation (10- 1), of

$$\phi P_{mmax} = 0.85 \ \phi P_o$$

In the case of a column with lateral ties, a minimum eccentricity of approximately $0.10h$ is specified which gives a maximum design axial load strength at zero eccentricity, in accordance with ACI Equation (10-2), of

$$\phi P_{mmax} = 0.80 \ \phi P_0 = 0.80 \times 1530 = 1224 \text{ kN}$$

References

1. Newnan, D.G., *Civil Engineering License Review*, Engineering Press, Austin, TX 1995
2. American Concrete Institute. *Building code requirements and commentary for reinforced concrete* (ACI 318-89), Detroit, MI, 1995.
3. American Concrete Institute. *Design handbook in accordance with the strength design method* Detroit, MI, 1985.
4. Ghosh, S. K. and Domel, A. W. *Design of concrete buildings for earthquake and wind forces.*Portland Cement Association, Skokie, IL, 1992.
5. Williams, A. *Design of Reinforced Concrete Structures.* Engineering Press, Austin,TX 1996.

Problems

4-1 to 4-4 Situation

The short column shown in the figure supports a steel column and base plate with the indicated loads. The short column is 450 mm square and is reinforced with Grade 400M deformed bars. Concrete compressive strength is14 MPa. The short column is subjected to axial load only and is effectively braced against sidesway by the floor slab.

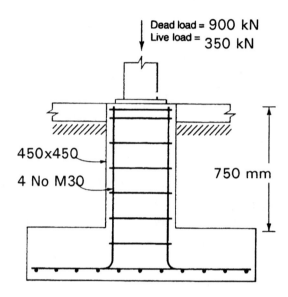

4-1. The design axial load strength of the short column is given most nearly by
(a) 1360 kN
(c) 1760 kN
(b) 1560 kN
(d) 1960 kN

4-2. The factored applied load is given most nearly by
(a) 1850 kN
(c) 1950 kN
(b) 1900 kN
(d) 2000 kN

4-3. The minimum allowable reinforcement area in the short column is most nearly
(a) 2000 mm²
(c) 2200 mm²
(b) 2100 mm²
(d) 2300 mm²

Solutions

4-1. For a column with lateral ties, the design axial load strength is given by Eq. (10-2) as:

$\phi P_n = 0.80\, \phi\, [0.85\, f'_c (A_g - A_{st}) + f_y A_{st}]$

$= 0.8 \times 0.7[0.85 \times 14(202{,}500 - 2800) + 400 \times 2800] / 1000$

$= 1958$ kN. The answer is (d).

4-2. The applied ultimate load is:

$P_u = 1.4 \times 900 + 1.7 \times 350$

$= 1855$ kN. The answer is (a).

4-3. The minimum allowable reinforcement area, in accordance with ACI Section, 10.9.1 is:

$\rho_{min} = 0.01\, A_g$

$= 0.01 \times 202{,}500$

$= 2025$ mm^2. The answer is (a).

Chapter 5
Waste Water and Solid Waste Treatment
Kenneth J. Williamson

Wastewater Flows

Wastewater flows are comprised of domestic and industrial wastewaters, infiltration, inflow, and storm water. Most modern sanitary sewers are separated from storm water systems so these flows are treated separately. Modern sewers are constructed so that inflow rates are assumed to be negligible.

Domestic flows are determined from water use rates with typical values being 0.57 m³ per day per capita. Industrial and infiltration flows vary widely. Specific data are required based upon industry and production rates.

Sewer Design

Hydraulics of Sewers

Sewers are designed as open channels, usually with a circular cross section. Flows in sewers are modeled using Manning's Eq. as:

(5-1)
$$V = \frac{1}{n} R^{2/3} S^{1/2}$$

where:
V = velocity (m/s);
n = Manning coefficient, 0.013;
R = hydraulic radius (m); and
S = slope of hydraulic grade line.

The relation between the hydraulic radius and the flow depth for a circular cross section is a complex relationship; as a result, flows in partially full sewers are calculated using nomographs as in Figure 5-1.

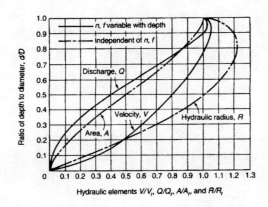

Figure 5-1.

Design calculations usually involve knowing the slope of the sewer (S), and the partially-full flow rate, Q. The nomograph is solved by assuming a pipe diameter (D), solving for the full flow rate, Q_f. From the ratio of Q/Q_f, the ratio of partial depth to pipe diameter (d/D) is obtained from the nomograph, from which the depth of flow is calculated. Sanitary sewers are designed to carry the peak flow with a depth of flow from one-half to full.

Wastewater Characteristics

Wastewater can be characterized by a large number of parameters. Typical concentration for design are given in Table 5-1; per capita loading rates for BOD_5 and TSS(Total Suspended Solids) are 100 and 120 g/capita-d, respectively.

Table 5-1. Typical Composition of Domestic Wastewater

Constituent	Concentration (mg/L)
Dissolved Solids	700
Volatile Dissolved Solids	300
Suspended Solids	220
Volatile Suspended Solids	135
5-day Biochemical Oxygen Demand	200
Organic Nitrogen	15
Ammonia Nitrogen	25
Total Phosphorus	8

Oxygen Demand

Theoretical Oxygen Demand

Theoretical oxygen demand is a value calculated as the oxygen required to convert organic compounds in the wastewater to carbon dioxide and water. To determine the ThOD, the chemical formula of the waste must be known.

Chemical Oxygen Demand

Chemical oxygen demand (COD) is an empirical parameter representing the equivalent amount of oxygen that would be used to oxidize organic compounds under strong chemical oxidizing conditions. The COD and the ThOD are approximately equal for most wastes.

Biochemical Oxygen Demand

Biochemical oxygen demand (BOD) is an empirical parameter representing the amount of oxygen that would be used to oxidize organic compounds by aerobic bacteria. The test involves seeding of diluted wastewater and measurement of the oxygen depletion after five days in 300 ml glass bottles. The 5-day BOD is related to the ultimate BOD as:

(5-2) $$BOD_s = BOD_L(1 - 10^{-kt})$$

where:
BOD_L = the maximum BOD exerted after a long time of incubation (mg/L); and
k = the base-10 BOD decay coefficient (d^{-1}).

The BOD test is accomplished at 20°C; the decay coefficient can be converted to other temperatures as:

(5-3) $$k_T = k_{20}\theta^{(T-20)}$$

where:
k_T, k_{20} = BOD decay coefficient at temperature T and 20°C, respectively; and
θ = temperature correction coefficient, 1.056 from 20-30°C and 1.135 from 4-20°C.

Nitrogenous Oxygen Demand

Nitrogenous oxygen demand results from the oxidation of ammonia to nitrate by nitrifying bacteria. The overall equation is:

(5-4) $$NH_4^+ + 2O_2 = NO_3^- + 2H^+ + 2H_2O$$

Nitrogenous oxygen demand is not included in either BOD_5 or COD values.

Wastewater Treatment

Wastewater treatment in the U.S. is managed under the Federal Water Pollution Control Act Amendments of 1972. Design of wastewater treatment facilities focuses on the two parameters of BOD and TSS. The general approach is to link together a series of unit processes to sequentially remove BOD and TSS to meet the discharge requirements. The unit processes are typically chosen to minimize treatment costs.

Process Analysis

Types of Reactions

Most reactions in wastewater treatment are consider to be homogeneous. In homogeneous reactions, the reaction occurs throughout the liquid and mass transfer effects can be ignored. Chlorination is an example of a homogeneous reaction. In heterogenous reaction, the reactants must be transferred to a reactive site and the products must be transferred away from the reactive site. Biological reactions in bacterial films is an example of a heterogenous reaction.

Reaction Rates

Chemical reactions have many different forms although the most common form is the conversion of a single product to a single reactant as:

(5-5) $$A = B$$

5-4

If the rate of the reaction is zero order, then the rate of conversion of A is:

(5-6)
$$\frac{dA}{dt} = -k_0$$

where:
k_0 = the zero order reaction coefficient (mg/L-d).

If the rate of the reaction is first order, then the rate of conversion of A is:

(5-7)
$$\frac{dA}{dt} = -k_1[A]$$

where:
k_1 = first order reaction coefficient (d^{-1}).

Reactor Types

Three important reactor types are batch, complete-mixed, and plug flow as shown in Figure 5-2. Batch reactors have no influent or effluent and have a hydraulic detention time of t_b which is the time between filling and emptying. Completely-mixed reactors have constant influents and effluent flow rates and are mixed so that spatial gradients of concentration are near zero. They have a hydraulic detention time as:

(5-8)
$$\theta_k = \frac{V}{Q}$$

where:
θ = hydraulic detention time (d);
V = reactor volume (L); and
Q = influent and effluent flow rate (L/d).

Batch

Plug flow reactors have constant influent and effluent flow rates, but lack internal mixing. The reactors tend to be long and narrow, and the transport of water occurs from one end to the other. Their hydraulic detention time is:

Complete mix

(5-9)
$$t_r = \frac{L}{v_e}$$

where:
t_r = hydraulic detention time (d);
L = length of reactor (m); and
v_e = liquid flow velocity (m/d).

Plug-flow

Figure 5-2

The application of mass balances to the three types of reactors assuming either zero- or first-order reaction rates results in the following equations for the effluent concentration of A at steady-state:

Batch Reactor, Zero-Order

(5-10)
$$A = A_0 - k_0 t$$

Complete-Mixed, Zero-Order
(5-11)
$$A = A_0 - k_0 \theta_h$$

Plug-Flow, Zero-Order

(5-12)
$$A = A_0 - k_0 t_r$$

Batch Reactor, First-Order

(5-13)
$$A = A_0 e^{(-k_1 t_b)}$$

Complete-Mixed, First-Order

(5-14)
$$A = \frac{A_0}{1 + k_1 \theta_h}$$

Plug-flow, First-Order

(5-15)
$$A = A_0 e^{-k_1 t_r}$$

Plug flow reactors will consistently give lower effluent concentrations as compared to complete-mixed reactors for all reaction rates greater than zero.

Physical Treatment Processes

Sedimentation

Sedimentation is divided into four types: discrete (Type 1); flocculent (Type 2); zone (Type 3); and compression (Type 4).

Discrete Sedimentation

Discrete sedimentation is described in Chapter 7. In discrete sedimentation, the removal is determined by the terminal settling velocity. Settling velocities for small Reynolds Numbers can be estimated by Stoke's Law as:

(5-16)
$$V_p = \frac{g(\rho_s - \rho_w)d^2}{18\mu}$$

where:
V_p = particle settling velocity (m/s);
g = gravitational constant (9.8 m/s^2);
ρ_p, ρ_w = density of particle and water, respectively (kg/m^3); and
μ = water viscosity (N-s/m^2).

For particles with V_p less than the overflow rate of the sedimentation basin, V_c, the removal, R, is:

(5-17)
$$R = \frac{V_p}{V_c}$$

R is 1.0 for all particles with $V_p > V_s$.

Flocculent Sedimentation

Flocculent sedimentation is similar to Type 1 settling except that the particle size increases as the particle settles in the sedimentation reactor. Type 2 or flocculent settling occurs in coagulation/flocculation treatment for water and in the top of secondary sedimentation basins.

Removal under flocculent sedimentation is determined empirically. Data must be collected to give isolines for percent removal as shown in Figure 5-3. Given such data, the total removal is:

(5-18)
$$R = \Sigma \left(\frac{\Delta h_n}{h_s} \right) \left(\frac{R_n + R_{n+1}}{2} \right)$$

where symbols are illustrated in Figure 5-3.

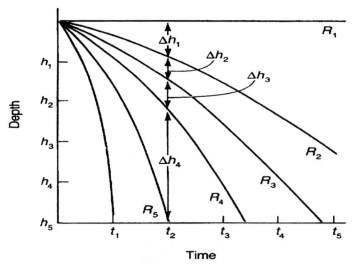

Figure 5-3. Empirical Analysis of Type 3 Settling

Zone Settling

Zone settling occurs in the bottom of a secondary sedimentation basin where the concentration of the solids increases to over 5000 mg/L. A schematic of a sedimentation basin coupled to a biological treatment reactor with cell recycle is shown in Figure 5-4. Under such conditions, the modeling of the zone sedimentation process is based upon the solids flux through the bottom section of the basin. The solids flux results from two components: gravity sedimentation and liquid velocity from the underflow (Eq. 5-19).

(5-19)
$$SF = SF_g + SF_r$$

where:

SF, SF_g, SF_r = solids flux, solids flux due to gravity,
and solids flux due to recycle, respectively (kg/m²-d).

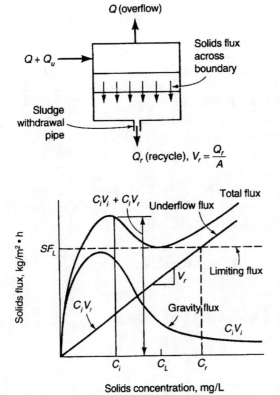

Figure 5-4. Schematic of Secondary Clarifier and Empirical Analysis of Type 3 Settling.

The resulting relationship is:

(5-20)
$$SF = C_i V_i + C_i V_r$$

where:

C_i = concentration of solids at a specified height in the basin (mg/L);
V_i = gravity settling velocity of sludge with concentration C_i (m/d);
V_r = downward fluid velocity from recycle (m/d),
   = Q_r / A_H
A_H = horizontal area of sedimentation basin.

Eq. 5-19 is graphed in Figure 5-4. Under a constant recycle flow and a defined sludge, the transport of solids through the bottom of the clarifier becomes limited by the limiting solids flux rate (SF_L) shown in Figure 5-4. Estimates of SF_L are obtained by graphical analysis illustrated in Figure 5-4. The horizontal area of the sedimentation basin is then calculated as:

(5-21)
$$A = \frac{(Q + Q_r)(C_{infl})}{SF_L}$$

or

(5-22)
$$A = \frac{Q_r C_r}{SF_L}$$

where:
C_{infl} = concentration of TSS in influent to sedimentation basin
C_r = concentration of TSS in recycle from sedimentation basin
Q, Q_r = plant influent and recycle flow rates.

Compression Sedimentation

Compression sedimentation involves the slow movement of water through the pores in a sludge cake. The process is modeled as an exponential decrease in height of the sludge cake as:

(5-23)
$$H_t - H_\infty = (H_0 - H_\infty)e^{-i(t - t_0)}$$

where:
H_t, H_∞, H_0 = sludge height after time t, a long period, and initially;
i = compression coefficient.

Biological Treatment Processes

Biological treatment involves the conversion of organic and inorganic compounds by bacteria with subsequent growth of the organisms. The common biological treatment processes in wastewater treatment involve the removal of various substrates including BOD, ammonia, nitrate, and organic sludges.

Description of Homogeneous Processes

The rate of removal of a substrate is given as:

(5-24)
$$r_{su} = -\frac{kXS}{(K_s + S)}$$

where:
r_{su} = rate of substrate removal (mg/L-d);
k = maximum substrate removal rate (mg/mg-d);
X = bacterial concentration, usually expressed as TVSS (mg/L);
S, K_s = substrate and half-velocity coefficient, respectively (mg/L).

The growth of organisms is given as:

(5-25)
$$r_X = Yr_{su} - k_d X$$

where:
r_X = rate of growth of bacteria (mg/L-d);
Y = substrate yield rate (mg/mg)
k_d = bacterial decay coefficient d^{-1}.

Complete-mixed Reactor without Cell Recycle

For the complete-mixed reactor without cell recycle (Figure 5-5a), the governing equations are:

(5-26)
$$X = \frac{Y(S_0 - S)}{(1 + k_d \theta_h)}$$

(5-27)
$$S = \frac{K_s(1 + \theta_h k_d)}{\theta_h(YK - k_d) - 1}$$

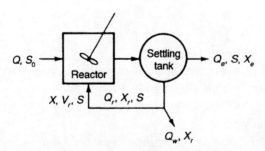

(a) Complete-Mix Without Cell Recycle

(b) Complete-Mix With Cell Recycle

Figure 5-5. Schematic of Complete-Mix Reactors

Completely-mixed Reactor with Cell Recycle

For the completely-mixed reactor with cell recycle (Figure 5-5b), the governing equation are:

(5-28)
$$\frac{1}{\theta_c} = Y(\frac{kSX}{S+K_s}) - k_d$$

where:
θ_c = solids retention time (d).

(5-29)
$$\frac{F}{M} = \frac{S_0}{\theta_h X}$$

where:
F/M = food to microorganism ratio (mg/mg-d).

(5-30)
$$X = \frac{\theta_c Y(S_0 - S)}{\theta_h(1 + k_d \theta_c)}$$

(5-31)
$$S = \frac{K_s(1 + \theta_c k_d)}{\theta_c(Yk - k_d) - 1}$$

Process design proceeds from selection of a solids retention time (θ_c) as the controlling parameter. The volume of the reactor is determined from Equation 5-30 with $\theta_h = V/Q$. The waste sludge flow rate is computed from:

(5-32)
$$\theta_c = \frac{VX}{Q_w X_w + Q_e X_e}$$

where:
Q_w, Q_e = sludge waste and effluent flow rates (L/d);
X_w = sludge waste concentration (X if from reaction, X_r if from the clarifier (mg/L); and
X_e = effluent TSS concentration, usually 20 mg/L (mg/L).

Plug Flow Reactor with Cell Recycle
The cell concentration in a plug flow reactor can be estimated using Equation 5-30. The effluent substrate concentration can be estimated from:

(5-33)
$$\frac{1}{\theta_c} = \frac{Yk(S_0 - S)}{(S_0 - S) + (1 + \frac{Q_r}{Q})K_s \ln(S_i/S)} - k_d$$

where:
S_i = the diluted substrate concentration entering the reactor (mg/L).

Description of Heterogeneous Processes

Trickling filters are designed using an empirical equation as:

(5-34)
$$A = Q\left(\frac{-\ln(S_{eff}/S_{infl})}{k_1 D}\right)^2$$

where:
A = horizontal area
k_1 = BOD decay coefficient;
D = filter depth;

Disinfection

Disinfection in wastewater treatment is almost universally accomplished using chlorine gas. The objective of disinfection is to kill bacteria, viruses, and amoebic cysts. Effluent standards for secondary treatment require fecal coliform levels of less than 200 and 400 per 100 ml for 30 day and 7 day averages, respectively.

Modeling of disinfection are described in Chapter 7-1. The major difference between chlorination of water supplies and wastewater is the reaction of chlorine with ammonia to produce chloramines as:

$$NH_3 + HOCl = NH_2Cl + H_2O$$

Chloramines can undergo further oxidation to nitrogen gas with their subsequent removal.

This stepwise oxidation of various compounds is shown in Figure 5-6. As chlorine is initially added, it reacts with easily oxidized compounds such as reduced iron, sulfur, and manganese. Further addition result in the formation of chloramines, collectively termed combined chlorine residual. Further addition of chlorine results in destruction of the chloramines, and ultimately to the formation of free chlorine residuals as HOCl and OCl⁻. The development of free chlorine residual is termed breakpoint chlorination.

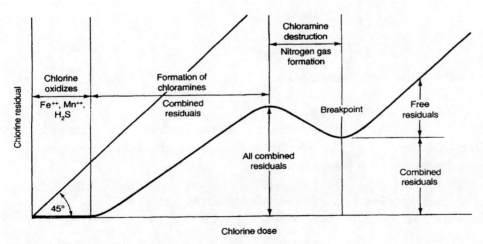

Figure 5-6. Chlorine Dosage versus Chlorine Residual

Problems

5.1
What is the required slope for a 0.3m diameter circular sewer with a flow of 0.014 m³/s and a minimum velocity of 0.6 m/s?

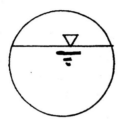

5.2
Compute the carbonaceous and nitroferrus oxygen demands for a waste with a chemical formula of $C_5H_7NO_2$

5.3
A lake with a volume of 5×10^6 m³ has a freshwater flow of 20 m³/s . A waste is dumped into the lake at a rate of 50 g/s with a decay rate of 0.2/d. What is the steady-state concentration? Assume lake is completely mixed.

5.4
A waste with a flow of 2.8 L/s is discharged to a small stream with a flow of 141 L/s. The waste has a 5-d BOD of 200 mg/L (K=0.2/d). What is the BOD after 1 day's travel in the stream?

5.5
A sedimentation basin has an overflow of 3 ft/hr. The influent wastewater has a particle distribution as follows:

Percent of particles	Settling Velocity (m/hr)
20	0.30 to 0.61
30	0.61 to 0.91
50	0.91 to 1.22

Determine the total removal in the sedimentation basin.

5.6
Disinfection with chlorine is known to be a first-order reaction. The first-order decay rate under a given concentration of chlorine is measured as 0.35/hr. The flow rate is 12000 L per hour and the desired removal of organism is from 10×10^6/100ml to less than 1/100ml.
Determine:
a) Volume of reactor required assuming the use of a completely-mixed reactor.
b) Volume of reactor required assuming the use of plug-flow reactor.

5.7

A Type III settling test was conducted on a waste activated sludge and the results were:

MLSS(mg/L)	Settling Velocity (m/hr)
4000	2.4
6000	1.2
8000	0.55
10,000	0.31
20,000	0.06

Determine the limiting solids flux if the concentration of the recycled solids is 10,000 mg/L.

5.8

Find the effluent soluble BOD and reactor cell concentration for an aerobic completely-mixed reactor with no recycle.

$K = 10$ mg/mg-d

$K_s = 50$ mg/l

$K_d = 0.10$/d

$Y = 0.6$ mg/mg

$S_0 = 200$ mg BOD/l

$\theta_h = \theta_c = 2$ d

Solutions

5.1

$$D_f = 0.30 \text{ m}$$

$$Q = 0.014 \text{ m}^3 / \text{s}$$

$$V = 0.6 \text{ m/s}$$

Need V / V_f or Q / Q_p or A / A_f or R / R_f

$$A_f = \frac{\pi D_f^2}{4} = 0.071 \text{ m}^2$$

$$A = \frac{Q}{V} = 0.023 \text{ m}^2$$

$$\frac{A}{A_f} = 0.32$$

From nomograph,

$$d / D_f = 0.35 \qquad d = 0.105 \text{ m}$$

$$R / R_f = 0.75$$

$$R_f = \frac{\pi D_f^2 / 4}{\pi D} = \frac{D_f}{4} = 0.075 \text{ m}$$

$$R = (0.75)(0.075) = 0.056 \text{ m}$$

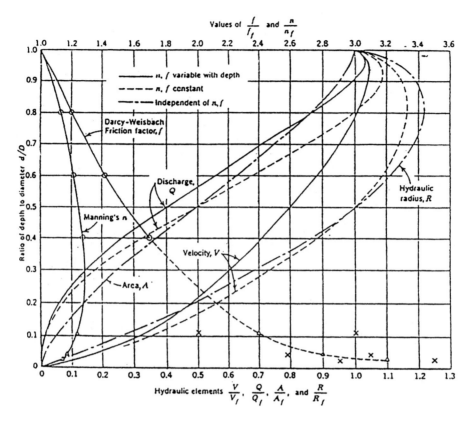

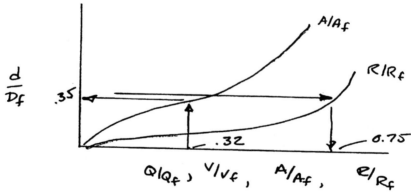

From Manning's Equation:

$$V = \frac{1}{n} R^{2/3} S^{1/2}$$

$$S = \left(\frac{nV}{1. R^{2/3}} \right)^2$$

$$= \left(\frac{(0.013)(0.6 m/s)}{(1.)(0.056)^{2/3}} \right)^2$$

$$= 0.0028 \text{ m/m}$$

5.2

For ThOD

$$C_5H_7NO_2 + 5O_2 \rightarrow 5CO_2 + NH_3 + 2H_2O$$

For NOD

$$NH_3 + 2O_2 \rightarrow HNO_3 + H_2O$$

For Carbonaceous oxygen demand

$$\frac{5 \text{moles } O_2}{1 \text{mole} C_5H_7NO_2} \times \frac{32 g O_2}{\text{mole} O_2} \times \frac{1 \text{mole} C_5H_7NO_2}{113 g} = \frac{1.4 g O_2}{g C_5H_7NO_2}$$

For Nitrogenous oxygen demand

$$\frac{2 \text{moles } O_2}{1 \text{mole} C_5H_7NO_2} \times \frac{32 g O_2}{\text{mole} O_2} \times \frac{1 \text{mole} C_5H_7NO_2}{113 g} = \frac{0.57 g O_2}{g C_5H_7NO_2}$$

5.3

$$C = \frac{S}{Q + KV}$$

$$= \frac{50 g/s}{20 m^3/s + (0.2/d)(5 \times 10^6 m^3)\left(\dfrac{1d}{8.64 \times 10^4 s}\right)}$$

$$= 1.58 g/m^3$$

5.4

$$BOD_L = \frac{BOD_5}{(1 + e^{-5K})}$$

$$= \frac{20 mg/L}{\left(1 - e^{-\left(5d \times \frac{.2}{d}\right)}\right)}$$

$$= 294 mg/L$$

After mixing

$$BOD_i = \frac{(294 mg/L)(2.8 L/s)}{141.6 L/s + 2.8 L/s}$$

$$= 5.76 mg/L$$

Assume Stream is plug flow

$$BOD = BOD_i e^{-Kt}$$

$$= (5.76 mg/L) e^{-(0.21d)(1d)}$$

$$= 4.7 mg/L$$

5.5

Particles	Avg V_p (cm/hr)	V_c (cm/hr)	R (%)
2.54-6.45	3.81	7.62	50
6.45-7.62	6.35	7.62	83
7.62-10.16	8.89	7.62	100

$$R_{total} = (0.2)(0.5) + (0.3)(0.83) + (0.5)(1.0)$$
$$= 0.85 \text{ or } 85\%$$

5.6

a).

$$\frac{A}{A_0} = \frac{1}{1 + K_1 \dfrac{v}{Q}} = \frac{1}{10 \times 10^6}$$

$$1 + K_1 \frac{v}{Q} = 1 \times 10^7$$

$$V = \frac{\left(1 \times 10^7\right)\left(12000\ell / hr\right)}{0.35 / hr}$$

$$= 3.4 \times 10^{11} \ell$$

b).

$$\frac{A}{A_0} = e^{-Kv/Q} = \frac{1}{10 \times 10^6}$$

$$-K \frac{v}{Q} = -16.1$$

$$V = \frac{-16.1\left(12000\ell / hr\right)}{0.35 / hr}$$

$$= 5.5 \times 10^5 \ell$$

5.7

MLSS (mg/l)	v_i (ft/hr)	$x_i v_i$ (mg-ft/l-hr)	v_i (cm/hr)	$x_i v_i$ (kg/m²-d)
4000	7.8	3.12×10^4	19.81	4.63
6000	3.8	2.28×10^4	9.65	3.11
8000	1.8	1.44×10^4	4.57	1.97
10,000	1.0	1.00×10^4	2.54	1.37
20,000	0.2	0.80×10^4	0.51	1.09

From graphical Solution,
limiting solids flux ≅ 7 Kg/m²-d

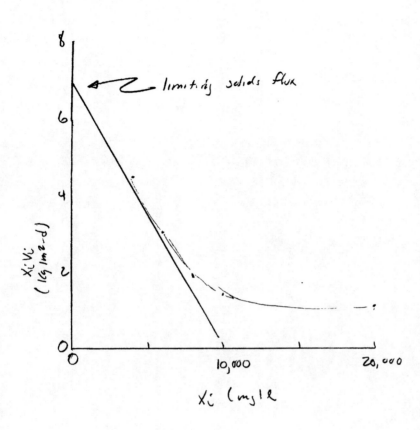

5.8

$$s = \frac{K_s\left(1 + \theta_h k_d\right)}{\theta_h\left(Y_k - k_d\right) - 1}$$

$$= \frac{(50\,mg\,/\,l)\left(1 + (2d)(0.10\,/\,d)\right)}{(2d)\left(0.6\,{}^{mg}\!/_{mg} \times 10\,{}^{mg}\!/_{mgd} - 0.10\,/\,d\right)^{-1}}$$

$$= 5.7\,mg\,/\,l$$

$$X = \frac{4(S_0 - S)}{1 + k_d \theta_h} = \frac{\left(0.6\,{}^{mg}\!/_{mg}\right)(200\,mg\,/\,l - 5.7\,mg\,/\,l)}{1 + (0.10\,/\,d)(2d)}$$

$$= 97\,mg\,/\,l$$

Chapter 6
TRANSPORTATION ENGINEERING
Robert W. Stokes

HIGHWAY CURVES

a. Simple (Circular) Horizontal Curves

The location of highway centerlines is initially laid out as a series of straight lines (tangent sections). These tangents are then joined by circular curves to allow for smooth vehicle operations at the design speed selected for the highway. The following figure shows the basic geometry of a simple circular curve. If the two tangents intersecting at the PI are laid out, and the angle Δ between them is measured, only one other element of the curve must be known to calculate the remaining elements. The radius of the curve (R) is the other element most commonly used.

Circular Curve Formulas

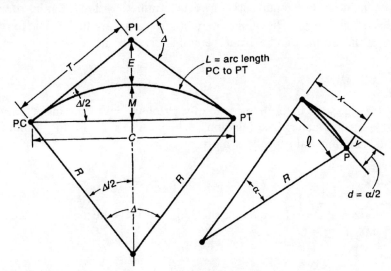

PC = Point of curvature (beginning of curve)
PI = Point of intersection
PT = Point of tangency (end of curve)
Δ = Intersection or central angle, degrees
R = Radius of curve, m
L = Length of curve, m
E = External distance, m
M = Middle ordinate, m
C = Long chord, m
ℓ = Length of arc between any two points on curve, m
α = Central angle subtended by arc ℓ, degrees
d = Deflection angle for any arc length ℓ, degrees
x = Distance along tangent from P.C. or P.T. to set any point P on curve, m
y = Offset (normal) from tangent at distance x to set any point P on curve, m

$L = \Delta R / 57.2958$
$T = R \tan \Delta/2$
$E = R(\sec \Delta/2 - 1) = R \text{ exsec } \Delta/2$
$M = R(1-\cos \Delta/2) = R \text{ vers } \Delta/2$
$C = 2R \sin \Delta/2$
$R = (C^2 + 4M^2) / 8M$ *

$\ell = (\alpha \times R) / 57.2958$
$d = \alpha/2 = 1718.873 \ \ell / R$ (in minutes)
    $= 28.64789 \ \ell / R$ (in degrees)
For any length x, $y = R - (R^2 - x^2)^{1/2}$
For any length ℓ, $X = R \sin \alpha$
                $Y = R (1-\cos \alpha) = R \text{ vers } \alpha$
* Formula for estimating R from field measurements

6-2

Example 6-1

Given the following horizontal curve data, determine the curve radius, R, length of curve, L, stationing (sta) of the PC, stationing of the PT, the long chord, C, and the deflection angle, d, of a point on the curve 30 m ahead of the PC.

PI = sta 2 + 170.00 (km)
$\Delta = 41°\ 10'$
T = 115.00 m

Solution

PC sta = PI sta - T = 2170.00 - 115.00 = 2 + 055.00
R = T/tan(Δ/2) = 115.00/0.3755 = 306.22 m
L = (ΔR)/57.2958 = 220.02 m
PT sta = PC sta + L = 2055.00 + 220.02 = 2275.02 m
C = 2R sin(Δ/2) = 2(306.22)(0.3516) = 215.32 m
d = (28.64789ℓ)/R = [(28.64689)(30)]/306.22 = 2.81°

The minimum radius of horizontal curvature is determined by the dynamics of vehicle operation and sight distance requirements. The minimum radius necessary for the vehicle to remain in equilibrium with respect to the incline of a horizontal curve is given by the following formula.

$R = V^2/[127 (e + f)]$

where V = vehicle speed in km/h, e = rate of superelevation ("banking") in m/m of width, and f = allowable side friction factor. The equation can be used to solve for the minimum radius for a given speed, or the maximum safe speed for a given radius.

Example 6-2

An existing horizontal curve has a radius of 235 m. What is the maximum safe speed on the curve. Assume e = 0.08 and f = 0.14.

Solution

$R = V^2/[127 (e + f)]$
$V = [R [127 (e + f)]]^{0.5}$
$V = [235 [127 (0.08 + 0.14]]^{0.5} = 81$ km/h

b. Vertical Curves

The vertical axis parabola (see following figure) is the geometric curve most commonly used in the design of vertical highway curves. The general equation of the parabola is

$Y = aX^2 + bX + c$

Where the constant a is an indication of the rate of change of slope, b is the slope of the back tangent (g_1), and c is the elevation of the PVC.

VERTICAL CURVE FORMULAS

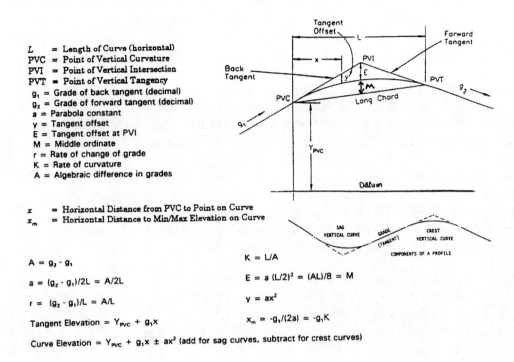

L = Length of Curve (horizontal)
PVC = Point of Vertical Curvature
PVI = Point of Vertical Intersection
PVT = Point of Vertical Tangency
g_1 = Grade of back tangent (decimal)
g_2 = Grade of forward tangent (decimal)
a = Parabola constant
y = Tangent offset
E = Tangent offset at PVI
M = Middle ordinate
r = Rate of change of grade
K = Rate of curvature
A = Algebraic difference in grades

x = Horizontal Distance from PVC to Point on Curve
x_m = Horizontal Distance to Min/Max Elevation on Curve

$$A = g_2 - g_1$$

$$a = (g_2 - g_1)/2L = A/2L$$

$$r = (g_2 - g_1)/L = A/L$$

$$\text{Tangent Elevation} = Y_{PVC} + g_1 x$$

$$K = L/A$$

$$E = a (L/2)^2 = (AL)/8 = M$$

$$y = ax^2$$

$$x_m = -g_1/(2a) = -g_1 K$$

Curve Elevation = $Y_{PVC} + g_1 x \pm ax^2$ (add for sag curves, subtract for crest curves)

Example 6-3

Determine the following for the sag vertical curve shown in the accompanying sketch:
a. Stationing and elevation of the PVC
b. Stationing and elevation of the PVT
c. Elevation of the curve 100 m from the PVC
d. Elevation of the curve 400 m from the PVC
e. Stationing and elevation of the low point of the curve

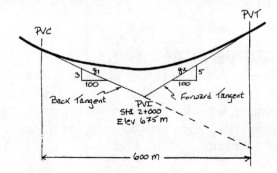

Solution

a. The curve is symmetric about the PVI, therefore:

  Sta PVC = Sta PVI - L/2 = 2000 - 600/2 = 1 + 700.0

  Elev of the PVC = Elev PVI + g_1(L/2) = 675 + 0.03(300) = 684.0 m

b. Sta PVT = Sta PVI + L/2 = 2000 + 600/3 = 2 + 300.0

  Elev of the PVT = Elev PVI + g_2(L/2) = 675 + 0.05(300) = 690.0 m

c. The elevation of the curve at any distance x from the PVC is equal to the tangent offset + the tangent elevation. To calculate the tangent offset, it is first necessary to evaluate the constant "a".

  a = (g_2 - g_1)/2L = [0.05 - (-0.03)]/[2(600)] = 0.000067

Tangent offset at x = 100 m from the PVC = ax^2 = 0.000067 $(100)^2$ = 0.67 m.

Tangent elevation 100 m from the PVC = PVC elev + g_1x = 684.0 m + (-0.03)100 = 681.0 m.

Curve elevation = tangent offset + Tangent elev = 0.67 + 681.0 = 681.67 m.

d. The elevation of the curve 400 m from the PVC can be calculated relative to the forward tangent or relative to the extension of the back tangent. Relative to the extension of the back tangent:

Tangent offset = ax^2 = 0.000067 $(400)^2$ = 10.67 m

Elev of the extended back tangent = PVC elev + g_1x = 684.0 + (-0.03)400 = 672.0 m

Curve elev = tangent offset + tangent elev = 10.67 + 672.0 = 682.67 m.

e. The low point of the curve occurs at a distance x_m from the PVC.

  x_m = -g_1/(2a) = -(-0.03)/[2(0.000067)] = 225 m (from the PVC).

Sta of the low point = Sta PVC + 225 = 1700 + 225 = 1 + 925.0

The tangent offset at 225 m from the PVC = ax^2 = 0.000067$(225)^2$ = 3.38 m.

The tangent elevation at this distance = PVC elev + g_1x = 684.0 + (-0.03)225 = 677.25 m.

The elevation of the low point = tangent offset + tangent elevation = 677.25 + 3.38 = 680.63 m.

c. Transition (Spiral) Curves

The spiral curve is used to allow for a transitional path from tangent to circular curve, from circular curve to tangent, or from one curve to another which have substantially different radii. The minimum length of spiral curve needed to achieve this transition can be computed from the following formula.

  $\ell_s = V^3/(46.7RC)$

where ℓ_s = minimum length of spiral (m), V = design speed (km/hr), R = circular curve radius (m), and C = rate of increase of centripetal acceleration (m/s³). The value C = 0.6 is generally used for highway curves.

TRANSITION (SPIRAL) CURVE FORMULAS

The basic geometry of the spiral curve is shown in the following sketch. The spiral is defined by its parameter "A" and the radius of the simple curve it joins. The product of the radius (r) at any point on the spiral and the corresponding spiral length (ℓ)from the beginning of the spiral to that point is equal to the product of the radius (R) of the simple curve it joins and the total length (ℓ_s) of the spiral as shown in the following equation.

  $r\ell = R\ell_s$ = constant = A^2

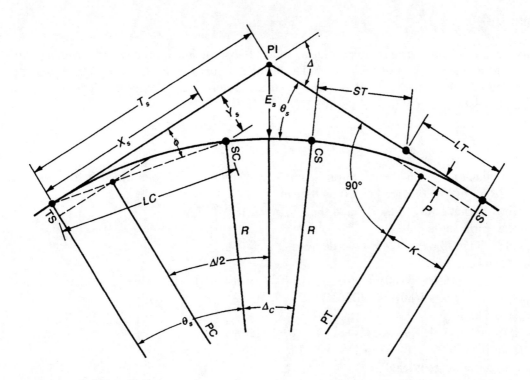

ℓ_s = total length of spiral from T.S. to S.C.

ℓ = spiral length from T.S. to any point on spiral

L_c = total length of circular curve

R = radius of circular curve

T_s = total tangent distance from the PI to the T.S. or S.T.

Δ = deflection angle between the tangents

θ_s = central angle of the entire spiral (spiral angle)

θ = central angle of any point on the spiral

Δ_c = central angle of simple curve

E_s = external distance

LC = long chord

LT = long tangent

ST = short tangent

X_s = tangent distance from T.S. to S.C.

x_s = tangent distance from T.S. to any point on the spiral

Y_s = tangent offset at the S.C.

y_s = tangent offset at any point on the spiral

k = simple curve coordinate (abscissa)

p = simple curve coordinate (ordinate)

ϕ = deflection angle at T.S. from the initial tangent to any point on the spiral

T.S. = tangent to spiral

S.C. = spiral to circular curve

C.S. = circular curve to spiral

S.T. = spiral to tangent

C_s = spiral deflection angle correction factor

$\theta_s = 28.648\,(\ell_s/R)$

$\theta = (\ell/\ell_s)^2\,\theta_s$

$\phi = (\theta/3) - C_s$

C_s (seconds) $= 0.0031\theta^3 + 0.00023\theta^5 \times 10^{-5}$

$Y_s = (\ell_s^2/6R) - (\ell_s^4/336R^3)$

$y_s = (\ell^2/6R) - (\ell^4/336R^3)$

$X_s = \ell_s - (\ell_s^3/40R^2)$

$p = Y_s - R(1 - \cos\theta_s)$

$k = X_s - R\sin\theta_s$

$T_s = (R + p)\tan(\Delta/2) + k$

$ST = Y_s/\sin\theta_s$

$LT = X_s - (Y_s/\tan\theta_s)$

(All distances in m and all angles in degrees unless noted otherwise)

Example 6-4

A transition spiral is being designed to provide a gradual transition into a circular curve with a radius of 435 m and a design speed of 90 km/hr. The PI of the curve is at Station 0+625.00 and the deflection angle between the tangents is 33.67 degrees. Determine the minimum length of spiral required and the stationing of the TS and the ST.

Solution

Assuming a value of C = 0.6, the minimum length of the spiral can be determined as follows.
$\ell_s = V^3/(46.7RC) = (90)^3/(46.7 \times 435 \times 0.6) = 60$ m
To determine the stationing of the TS and the ST, proceed as follows.
Compute the spiral angle.
$\theta_s = 28.648 (\ell_s/R) = 28.648 (60/435) = 4$ degrees
Compute Y_s and X_s.
$Y_s = (\ell_s^2/6R) - (\ell_s^4/336R^3) = [(60)^2/(6)(435)] - [(60)^4/(336)(435)^3] = 1.379$ m
$X_s = \ell_s - (\ell_s^3/40R^2) = 60 - [(60)^3/(40)(435)^2] = 59.972$ m
Compute p and k.
$p = Y_s - R(1 - \cos \theta_s) = 1.379 - 435(1 - \cos 4) = 0.319$ m
$k = X_s - R \sin \theta_s = 59.972 - (435)\sin 4 = 29.628$ m
Compute the spiral tangent length.
$T_s = (R + p) \tan (\Delta/2) + k = (435 + 0.319)\tan (33.67/2) + 29.628 = 161.349$ m
Compute the central angle of the circular curve.
$\Delta_c = \Delta - 2\theta_s = 33.67 - 2(4) = 25.67$ degrees
Compute the length of the circular curve.
$L = (\Delta_c R)/57.2958 = [25.67(435)]/57.2958 = 194.891$ m
Compute the stationing of the TS.
TS Sta = PI Sta - T_s = 0+625 - 0+161.349 = 0+463.651
Compute the stationing of the ST.
ST Sta = TS Sta + L + $2\ell_s$ = 0+463.651 + 0+194.891 + 2(60) = 0+778.542

SIGHT DISTANCE

The ability of drivers to see the road ahead is of utmost importance in the design of highways. This ability to see is referred to as sight distance and is defined as the length of highway ahead that is visible to the driver.

Stopping sight distance is the sum of the distance traveled during the perception-reaction time and the distance traveled while braking to a stop. The stopping sight distance can be determined from the following equation.
$$S = 0.278Vt + V^2/[254 (f \pm g)]$$
Where S = stopping sight distance (m), V = vehicle speed (km/hr), t = perception-reaction time (assumed to be 2.5 sec), f = coefficient of friction, and g = grade (*plus* for uphill, *minus* for downhill), expressed as a decimal.

The required middle ordinates (M) for clear sight areas to satisfy stopping sight distance (S) requirements as a function of the radii (R) of **simple horizontal curves** can be determined using the following equation. The stopping sight distance is measured along the centerline of the inside lane of the curve.

$$M = R[1 - \cos(28.65 \; S/R)]$$

Example 6-5

A large outcropping of rock is located at M (middle ordinate) = 10 m from the centerline of the inside lane of a proposed highway curve. The radius of the proposed curve is 120 m. What speed limit would you recommend for this curve? Explain your answer. Assume $f = 0.28$, $e = 0.10$, perception-reaction time = 2.5 sec., and level grade.

Solution

To solve this problem you must determine whether sight distance or the radius of the curve controls the speed.

Determine the available sight distance.

$$M = R[1 - \cos(28.65 \; S/R)]$$
$$S = (R/28.65)\cos^{-1}[(R - M)/R] = (120/28.65)\cos^{-1}[(120-10)/120] = 98.67 \text{ m}.$$

Determine the maximum safe speed on the curve as a function of stopping sight distance.

$$S = 0.278Vt + V^2/[254(f \pm g)]$$
$$98.67 = 0.278V(2.5) + V^2/[254(0.28)]$$
$$V^2 + 49.43V - 7017.41 = 0$$

and from the quadratic equation

$$V = 62.6 \text{ km/hr}$$

Determine the maximum safe speed on the curve as a function of the radius.

$$R = V^2/[127(e + f)]$$
$$V = [127R(e + f)]^{0.5} = [127(120)(0.10 + 0.28)]^{0.5} = 76.1 \text{ km/hr}$$

Because the sight distance is sufficient only for a speed of about 62.6 km/hr, set speed limit at 60 km/hr.

Formulas for **stopping and passing sight distances** for **vertical curves** are summarized in the following sketch. The formulas are based on an assumed driver eye height (h_1) of 1070 mm. The height of object (h_2) above the roadway surface is assumed to be 150 mm for stopping sight distance and 1300 mm for passing sight distance.

SIGHT DISTANCE ON VERTICAL CURVES

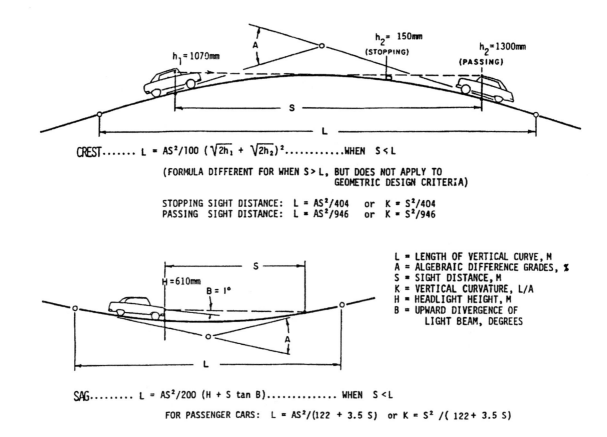

CREST....... $L = AS^2/100 \; (\sqrt{2h_1} + \sqrt{2h_2})^2$............WHEN S < L

(FORMULA DIFFERENT FOR WHEN S > L, BUT DOES NOT APPLY TO
GEOMETRIC DESIGN CRITERIA)

STOPPING SIGHT DISTANCE: $L = AS^2/404$ or $K = S^2/404$
PASSING   SIGHT DISTANCE: $L = AS^2/946$ or $K = S^2/946$

L = LENGTH OF VERTICAL CURVE, M
A = ALGEBRAIC DIFFERENCE GRADES, %
S = SIGHT DISTANCE, M
K = VERTICAL CURVATURE, L/A
H = HEADLIGHT HEIGHT, M
B = UPWARD DIVERGENCE OF
     LIGHT BEAM, DEGREES

SAG........ $L = AS^2/200 \; (H + S \tan B)$.............. WHEN S < L

FOR PASSENGER CARS: $L = AS^2/(122 + 3.5 S)$ or $K = S^2/(122 + 3.5 S)$

Example 6-6

A crest vertical curve joins a +2% grade with a -2% grade. If the design speed of the highway is 95 km/hr, determine the minimum length of curve required. Assume f = 0.29 and the perception-reaction time = 2.5 sec. Also assume S < L.

Solution

Determine the stopping sight distance required for the design conditions.
$S = 0.278Vt + V^2/[254 \, (f \pm g)] = 0.278 \, (95)(2.5) + (95)^2/[254(0.29 - 0.02)] = 197.62$ m
(Note that the worst case value for g is used.)
Determine the minimum length of vertical curve to satisfy the required sight distance.
$L = AS^2/404 = [4(197.62)^2]/404 = 386.67$ m

TRAFFIC CHARACTERISTICS

The traffic on a roadway may be described by three general parameters: volume or rate of flow, speed, and density.

Traffic volume is the number of vehicles that pass a point on a highway during a specified time interval. Volumes can be expressed in daily or hourly volumes or in terms of subhourly rates of flow. There are four commonly used measures of daily volume.

1) **Average Annual Daily Traffic** (AADT) is the average 24-hour traffic volume at a specific location over a full year (365 days).

2) **Average Annual Weekday Traffic** (AAWT) is the average 24-hour traffic volume occurring on weekdays at a specific location over a full year.

3) **Average Daily Traffic** (ADT) is basically an estimate of AADT based on a time period less than a full year.

4) **Average Weekday Traffic** (AWT) is an estimate of AAWT based on a time period less than a full year.

Highways are generally designed on the basis of the **directional design hourly volume (DDHV)**.

$$DDHV = AADT \times K \times D$$

where AADT = average annual daily traffic (veh/day), K = proportion of daily traffic occurring in the design hour, D = proportion of design hour traffic traveling in the peak direction of travel.

For design purposes, K often represents the proportion of AADT occurring during the **thirtieth highest hour** of the year on rural highways, and the fiftieth highest hour of the year on urban highways.

Traffic volumes can also exhibit considerable variation within a given hour. The relationship between hourly volume and the maximum 15-minute rate of flow within the hour is defined as the **peak-hour factor (PHF)**.

$$PHF = V_h/(4V_{15})$$

where V_h = hourly volume (vph), V_{15} = maximum 15-minute rate of flow within the hour (veh).

The second major traffic stream parameter is speed. Two different measures of average speed are commonly used.

Time mean speed is the average speed of all vehicles passing a point on the highway over a given time period. **Space mean speed** is the average speed obtained by measuring the instantaneous speeds of all vehicles on a section of roadway. Both of these measures can be calculated from a series of measured travel times over a measured distance from the following equations

$$\mu_t = (\Sigma(d/t_i))/n$$
$$\mu_s = (nd)/\Sigma t_i$$

where μ_t = time mean speed (m/sec or km/hr), μ_s = space mean speed (m/sec or km/hr), d = distance traversed (m or km), n = number of travel times observed, t_i = travel time of ith vehicle (sec or hr)

Because the space mean speed weights slower vehicles more heavily than time mean speed, it results in a lower average speed. The relationship between these two mean speeds is

$$\mu_t = \mu_s + \sigma_s^2/\mu_s$$

where σ_s^2 = the variance of the space speed distribution.

Example 6-7

The following travel times were observed for four vehicles traversing a 1-km segment of highway.

Vehicle	Time (minutes)
1	1.6
2	1.2
3	1.5
4	1.7

Calculate the space and time mean speeds of these vehicles.

Solution
The space mean speed is
$$\mu_s = (nd)/\Sigma t_i = 4(1)/(1.6 + 1.2 + 1.5 + 1.7) = 0.67 \text{ km/minute} = 40.0 \text{ km/hr.}$$
The time mean speed is
$$\mu_t = \Sigma(d/t_i)/n = [(1/1.6) + (1/1.2) + (1/1.5) + (1/1.7)]/4 = 0.68 \text{ km/minute} = 40.8 \text{ km/hr.}$$

The third measure of traffic stream conditions, **density** (sometimes referred to as concentration), is the number of vehicles traveling over a unit length of highway at a given instant in time.

The general equation relating flow, density and speed is
$$q = k\mu_s$$
where q = rate of flow (vph), μ_s = space mean speed (km/hr) and k = density (veh/km)

The relationship between density and flow shown in the following figure is commonly referred to as the "fundamental diagram of traffic flow." Note in the figure that 1) when density is zero, the flow is also zero, 2) as density increases, the flow also increases, and 3) when density reaches its maximum, referred to as jam density (k_j), the flow is zero.

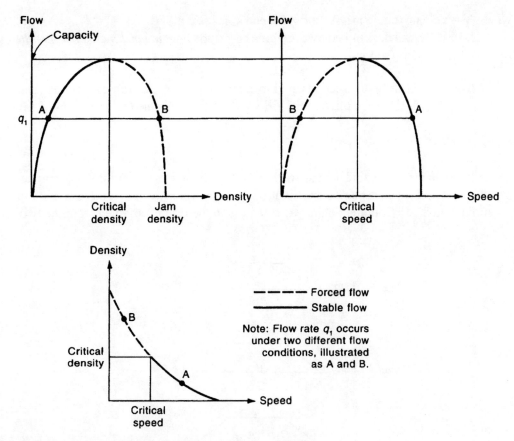

Relationships among Speed, Density and Rate of Flow

EARTHWORK

One of the major objectives in evaluating alternative route locations is to minimize the amount of cut and fill. A common method of determining the volume of earthwork is the **average end area** method. This method is based on the assumption that the volume between two consecutive cross sections is the average of their areas multiplied by the distance between them.

$$V = (L/54)(A_1 + A_2)$$

where V = volume (cubic meters), A_1 and A_2 = end areas (sq. m.)

L = distance between cross sections (m)

In situations where there is a significant difference between A_1 and A_2, it may be advisable to calculate the volume as a pyramid

$$V = (1/81)(\text{area of base})(\text{length})$$

where V is in cubic meters, the area of the base is in square meters, and the length is in meters.

The average end area and the pyramid methods provide "reasonably accurate" estimates of volumes of earthwork. When a more precise estimate of volume is desired, the **prismoidal formula** is frequently used.

$$V = (L/6)(A_1 + 4A_m + A_2)$$

where V = volume (cubic m), A_1 and A_2 = end areas (sq. m)

A_m = middle area determined by averaging corresponding *linear dimensions (not the end areas)* of the end sections (sq. m)

When materials from cut sections are moved to fill sections, **shrinkage factors** (generally in the range of 1.10 to 1.25) are applied to the fill volumes to determine the quantities of fill required.

Example 6-8

Given the following trapezoidal cross sections for a temporary access ramp. Determine the volume in cubic meters between stations 4+320 and 4+400 by 1) the average end area method, and 2) the prismoidal method. Comment on any differences in the results obtained from the two methods.

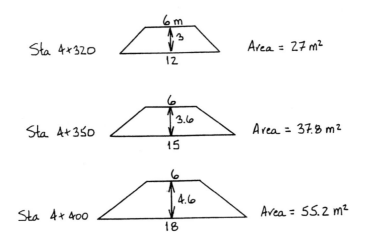

Solution

1. For the average end area method:

$$V = L/2 \ (A_1 + A_2)$$

The total volume is the sum of the volumes between stations 4+320 and 4+350 and between 4+350 and 4+400.

$$V = (30/2)(27.0 + 37.8) + (50/2)(37.8 + 55.2) = 3,297.9 \ m^3$$

2. For the prismoidal method:

$$V = (L/6)(A_1 + 4A_m + A_2)$$

The middle areas (A_m) are calculated from the averages of the base, top and height dimensions of the trapezoids.

The middle area between stations 4+320 and 4+350 is [(6.0 + 13.5)/2]x3.3 = 32.2 sq. m. The middle area between stations 4+350 and 4+400 is [(6.0 + 16.5)/2]x4.1 = 46.1 sq. m.

The total volume is the sum of the volumes between stations 4+320 and 4+350 and between 4+350 and 4+400.

$$V = (30/6)[27.0 + 4(32.2) + 37.8] + (50/6)[37.8 + 4(46.1) + 55.2] = 3,279.7 \ m^3$$

Comment: The prismoidal volumes are theoretically more precise than the average end area volumes. The average end area method tends to overestimate volumes, as in this case.

REFERENCES

American Association of State Highway and Transportation
Officials (AASHTO) . *A Policy on Geometric Design of Highways and
Streets,* Washington, DC, 1994.

Garber N.J. and L.A. Hoel. *Traffic and Highway Engineering,*
West Publishing Co., St. Paul, MN, 1988.

Hickerson, T.F. *Route Location and Design,* 5th ed., McGraw-Hill
NY.

Institute of Transportation Engineers (ITE). *Traffic Engineering
Handbook,* 4th ed., Washington, DC, 1992.

Newnan, D.G., Editor. *Civil Engineering License Review*
12th Ed., Engineering Press, Austin,TX 1995.

Transportation Research Board (TRB) . *Highway Capacity Manual,*
Special Report 209, 3rd Edition, TRB, Washington, DC, 1994.

PROBLEMS

6-1. For a simple horizontal curve with Δ = 12 degrees, R = 400 m and PI at Station 0+241.782, the stationing of the PC and PT, respectively, would be most nearly:

    (A)   0+199.7, 0+283.8
    (B)   0+199.7, 0+283.5
    (C)   0+158.0, 0+283.8
    (D)   0+195.6, 0+279.4

6-2. The minimum radius (m) for a simple horizontal curve with a design speed of 110 km/hr is most nearly: (assume e = 0.08 and f = 0.14)

    (A)   425
    (B)   395
    (C)   550
    (D)   435

6-3. A vehicle hits a bridge abutment at a speed estimated by investigators as 25 km/h. Skid marks of 30 m on the pavement (coefficient of friction, f = 0.35) followed by skid marks of 60 m on the gravel shoulder (f = 0.50) approaching the abutment are observed at the accident site. The grade is level. The initial speed (km/h) of the vehicle was at least:

    (A)   105
    (B)   90
    (C)   110
    (D)   95

Questions 4-5

A +3.9% grade intersects a +1.1% grade at station 0 + 625.0 and elevation 305.0 m.

6-4. The minimum length (m) of the vertical curve for a design speed of 80 km/h is most nearly: (assume perception-reaction time = 2.5 sec, f = 0.30, and S<L)
    (A)   85
    (B)   130
    (C)   55
    (D)   110

6-5. The elevation (m) of the middle point of the curve is most nearly:

    (A)   305.00
    (B)   305.45
    (C)   304.45
    (D)   304.55

SOLUTIONS

6-1.

Calculate the tangent length
$T = R \tan(\Delta/2) = 400 \tan(12/2) = 42.042$ m.
Calculate the length of the curve
$L = (R\Delta)/57.2958 = 83.776$ m
PC sta = PI sta - T = 0+241.782 - 0+042.042 = 0+199.740
PT sta = PC sta + L = 0+199.740 + 0+083.776 = 0+283.516
ANSWER = B

6-2.

$R = V^2/[127(e + f)] = (110)^2/[127(0.08 + 0.14)] = 433$ m
ANSWER = D

6-3.

The only known speed is the final collision speed of 25 km/h. Therefore, consider the braking distance on the gravel first.
Braking distance = $(V_0^2 - V^2)/[254(f \pm g)]$, where V_0 and V represent the initial and final speeds, respectively.
Braking distance (gravel) = 60 m = $(V_0^2 - 25^2)/(254)(0.5)$
Solving for V_0, $V_0 = 90.8$ km/h = speed at the beginning of the gravel and at the end of the pavement skid.
Therefore, for the pavement skid,
Braking distance (pavement) = 30 m = $(V_0^2 - 90.8^2)/(254)(0.35)$
Solving for V_0, $V_0 = 104.5$ km/h
ANSWER = A

Comment: Note that this solution does not account for any speed reduction prior to the beginning of the skid marks on the pavement. Therefore, we conclude that the vehicle speed was <u>at least</u> 104.5 km/h.

6-4.

The required stopping sight distance is:
   $S = 0.278Vt + V^2/[254(f \pm g)] = 0.278(80)(2.5) + (80)^2/[254(0.30 + 0.011)] = 136.62$ m
The minimum length of crest vertical curve to satisfy the required stopping sight distance is:
   $L = (|A|S^2)/404 = [|(1.1 - 3.9)|(136.62)^2]/404 = 129.36$ m
ANSWER = B

6-5.

The midpoint offset is:
   $E = (AL)/8 = [(0.011 - 0.039)(129.36)]/8 = -0.45$ m

Therefore, the elevation of the midpoint of the curve = elevation of the PVI (305 m) - 0.45 m = 304.55 m

ANSWER = D

Note: The elevation of the midpoint could also be found using the basic properties of the parabola. The offset at the midpoint of the curve $(x = L/2)$ is:

$$y = ax^2 = [(g_2 - g_1)/2L](L/2)^2 = [(0.011 - 0.039)/2(129.36)]/(129.36/2)^2 = -0.45 \text{ m}$$

Chapter 7
Water Purification and Treatment
Kenneth J. Williamson

Water Distribution
Water distribution systems involve a water source, a transmission line, a distribution network, pumping, and storage.

Water Source
The importance of the water source is primarily related to water elevation and water quality. The water elevation will determine the pumping required to maintain adequate flows in the network.

Water quality is largely determined by the source. Typical water sources include reservoirs, lakes, rivers, and underground aquifers. Surface water sources typically have low dissolved solids, but high suspended solids. As a result, treatment of such sources requires coagulation followed by sedimentation and filtration to remove suspended solids. Groundwater sources are low in suspended solids, but may be high in dissolved solids and reduced compounds. Groundwater, if high in dissolved solids, requires chemical precipitation or ion exchange for removal of dissolved ions.

A transmission line is required to convey water from the sources and storage to the distribution the system. Arterial main lines supply water to the various loops in the distribution system. The main lines are arranged in loops or in parallel to allow for repairs. Such lines are designed using the Hazen-Williams formula for single pipes (Eq. 7-1) and the equivalent pipe method for single loops. Multiple loops are designed using Hardy Cross methodology.

$$v = kCr^{0.63}s^{0.54}$$

(7-1)

where:
v = pipe velocity (m/s),
k = constant (2.79),
C = roughness coefficient (100 for cast iron pipe),
r = hydraulic radius (m), and
s = slope of hydraulic grade line (m/m).

The equivalent pipe method involves two procedures: conversion of pipes of unequal diameters in series into a single length of pipe, and conversion of pipes in parallel flow from the same nodes into a single pipe with a single diameter. The headloss is maintained through each conversion.

The Distribution Network
The distribution network is comprised of the arterial mains, distribution mains, and smaller distribution piping. The controlling design variable for water distribution networks is the pressure under maximum flow with a minimum value of 140 to 280 KPa and a maximum value of 690 KPa.

Flow for the water distribution network is calculated as the sum of domestic, irrigation, industrial and commercial, and fire requirements. The various flows for a given community are typically obtained from historical data of water use. Without such data, estimates can be made from average values listed in Table 7-1.

Table 7-1. Water Flow Rates

Flow	Method of Estimation
Average domestic flow	Population served, 0.4 m³/cap-d, 3,000 to 10,000 persons/km²
Irrigation flow	Maximum of 0.75 times average daily flow for arid climates
Industrial/commercial flow	Computed from known industries and area of commercial districts

Water Storage

Water storage is required to maintain pressure in the system, to minimize pumping costs, and meet emergency demands. Typically the storage is placed so that the load center or distribution network is between the water source and the storage as shown in Figure 7-1.

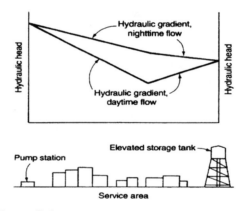

Figure 7-1

Storage requirements are calculated based upon variations in hourly flow and fire requirements. If the pumping capacity is set at some value less that the maximum hourly flow, then all flow requirements above that value will have to be provided by storage.

Pumping Requirements

The pumping system needs adequate capacity for design flows taking into account the supply from storage. The required head must be adequate to maintain 140 KPa at the load center during maximum flows.

Water Quality

The design of water treatment facilities requires calculations of concentrations and masses of various constituents in water and chemical additions. Common elements and radicals of various constituents are listed in Table 7-2, and common water treatment chemicals are listed in Table 7-3. The equivalent weight is equal to the molecular weight divided by the electrical charge.

Table 7-2. Common Elements and Radicals

Name	Symbol	Atomic Weight	Valence	Equivalent Weight
Aluminum	Al	27.0	+3	9.0
Calcium	Ca	40.1	+2	20.0
Carbon	C	12.0	-4	
Chlorine	Cl	35.5	-1	35.5
Fluorine	F	19.0	-1	19.0
Hydrogen	H	1.0	+1	1.0
Iodine	I	126.9	-1	126.9
Iron	Fe	55.8	+2	27.9
			+3	
Magnesium	Mg	24.3	+2	12.15
			+4	
			+7	
Nitrogen	N	14.0	-3	
			+5	
Oxygen	O	16.0	-2	8.0
Potassium	K	39.1	+1	39.1
Sodium	Na	23.0	+1	23.0
Ammonium	NH_4^+	18	+1	18.0
Hydroxyl	OH^-	17.0	-1	17.0
Bicarbonate	HCO_3^-	61.0	-1	61.0
Carbonate	$CO_3^=$	60.0	-2	30.0
Nitrate	NO_3^-	46.0	-1	46.0
Hypochlorite	OCl^-	51.5	-1	51.5

Table 7-3. Common Inorganic Chemicals for Water Treatment

Name	Formula	Usage	Molecular Weight	Equivalent Weight
Activated Carbon	C	Taste and Odor	12.0	
Aluminum Sulfate	$Al_2(SO_4)_3*14.3H_2O$	Coagulation	600	100
Ammonia	NH_3	Chloramines, disinf.	17.0	
Ammonium fluosilicate	$(NH_4)_2SO_4$	Fluoridation	178	
Calcium carbonate	$CaCO_3$	Corrosion control	132	66.1
Calcium fluoride	CaF_2	Fluoridation	78.1	
Calcium hydroxide	$Ca(OH)_2$	Softening	74.1	37.0
Calcium hypochlorite	$Ca(ClO)_2*2H_2O$	Disinfection	179	
Calcium oxide	CaO	Softening	56.1	28.0
Carbon dioxide	CO_2	Recarbonation	44.0	22.0
Chlorine	Cl_2	Disinfection	71.0	
Chlorine dioxide	ClO_2	Taste and odor	67.0	
Ferric chloride	$FeCl_3$	Coagulation	162	54.1
Ferric hydroxide	$Fe(OH)_3$		107	35.6
Fluorosilicic acid	H_2SiF_6	Fluoridation	144	
Oxygen	O_2	Aeration	2.0	16.0
Sodium bicarbonate	$NaHCO_3$	pH adjustment	84.0	84.0
Sodium carbonate	Na_2CO_3	Softening	106	53.0
Sodium hydroxide	NaOH	pH adjustment	40.0	40.0
Sodium hypochlorite	NaClO	Disinfection	74.4	
Sodium fluosilicate	Na_2SiF_6	Fluoridation	188	

Ion Balances

Electroneutrality requires that water has an equal number of equivalents of cations and anions. Often bar graphs are used to show this relationship as shown in Figure 7-2. From such diagrams, concentrations of alkalinity, carbonate (calcium and magnesium associated with alkalinity) and non-carbonate hardness (calcium and magnesium in other forms) can be easily identified.

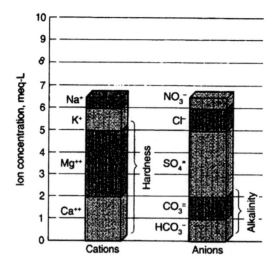

Figure 7-2

General Chemistry Concepts

A number of concepts from general chemistry are required for engineering calculations related to water treatment.

Oxidation-Reduction Reactions

Oxidation-reduction reactions involve the transfer of electrons from an electron donor to an electron acceptor. The easiest method to construct balance redox reaction is through the use of half-reactions. A list of common half-reactions related to water treatment are in Table 7-4.

Table 7-4 Common Half-Reactions for Water Treatment

Reduced Element	Half-Reaction
Cl	$1/2Cl_2 + e^- = Cl^-$
Cl	$1/2ClO^- + H^+ + e^- = 1/2Cl^- + 1/2H_2O$
Cl	$1/8ClO_4^- + H^+ e^- = 1/8Cl^- + 1/2H_2O$
Fe	$1/2Fe^{+2} + e^- = 1/2Fe$
Fe	$Fe^{+3} + e^- = Fe^{+2}$
Fe	$1/3Fe^{+3} + e^- = 1/3Fe$
I	$1/2I_2 + e^- = I^-$
N	$1/8NO_3^- + 5/4H^+ + e^- = 1/8NH_4^+ + 3/8H_2O$
N	$1/5NO_3^- + 6/5H^+ + e^- = 1/10N_2 + 3/5H_2O$
O	$1/4O_2 + H^+ + e^- = 1/2H_2O$

Henry's Law

Henry's Law states that the weight of any dissolved gas is proportional to the pressure of the gas.

$$C_{equil} = \alpha\, p_{gas} \tag{7-2}$$

where:
C_{equil} = equilibrium dissolved gas concentration,
P_{gas} = partial pressure of gas above liquid, and
α = Henry's Law constant.

Equilibrium Relationships

For an equilibrium chemical equation expressed as:

$$A + B = C + D \tag{7-3}$$

the relationship of concentrations at equilibrium can be approximated as:

$$K_{eq} = \frac{[C][D]}{[A][B]} \tag{7-4}$$

where:
K_{eq} = equilibrium constant, and
[ ] = molar concentrations.
Some commonly used equilibrium constants are listed in Table 7-5.

Table 7-5. Common Equilibrium Constants

Equation	K_{eq}
$H_2CO_3^= = H^+ + HCO_3^-$	$10^{-6.4}$
$HCO_3^- = H^+ + CO_3^=$	$10^{-10.3}$
$NH_3 + H_2O = NH_4^+ + OH^-$	$10^{-4.7}$
$CaOH^+ = Ca^{+2} + OH^-$	$10^{-1.5}$
$MgOH^+ = Mg^{+2} + OH^-$	$10^{-2.6}$
$HOCl = H^+ + OCl^-$	$10^{-7.5}$

A common equilibrium relationship is the disassociation of water which is given as:

$$H_2O = H^+ + (OH)^- \tag{7-5}$$

which has a K_{eq} value of 10^{-7}. The hydrogen ion concentration is typically represented by the pH value or the negative log of its concentration as:

$$pH = -\log(H^+) \tag{7-6}$$

Alkalinity

Alkalinity is a measure of the ability of water to consume a strong acid. In the alkalinity test, water is titrated with a strong acid to pH 6.4 and to pH 4.5. The amount of acid that is used to reach the first pH endpoint is termed the carbonate alkalinity, and to reach the second end point is the total alkalinity. Alkalinity is expressed as calcium carbonate.

Alkalinity is the sum of bicarbonate, carbonate, hydroxide minus hydrogen ion. For most sources to be used for drinking water, alkalinity can be approximated as either carbonate and bicarbonate. Carbonate alkalinity is the predominate species when the pH is above 10.8, and bicarbonate is the predominate species when the pH is between 6.9 and 9.8.

Solubility Relationships

The equilibrium between a compound in its solid crystalline state and its ionic form in solution where:

$$X_a Y_b = aX^{b+} + bY^{a-}$$

$$(7-7)$$

is given by:

$$K_{sp} = \left[X^{b+}\right]^a \left[Y^{a-}\right]^b$$

$$(7-8)$$

where:
K_{sp} = solubility product.

Solubility products for common precipitation reactions in water treatment are given in Table 7-6.

Table 7-6. Common Solubility Products

Compound	K_{sp}
Magnesium carbonate	4×10^{-5}
Magnesium hydroxide	9×10^{-12}
Calcium carbonate	5×10^{-9}
Calcium hydroxide	8×10^{-6}
Aluminum hydroxide	1×10^{-32}
Ferric hydroxide	6×10^{-38}
Ferrous hydroxide	5×10^{-15}
Calcium fluoride	3×10^{-11}

Dissolved Oxygen Relationships

The dissolved oxygen concentration in a stream is determined by the rate of oxygen consumption, commonly caused by the discharge of oxygen demanding substances, and the rate of reoxygenation from the atmosphere.

Biochemical Oxygen Demand

The concentration of oxygen demanding substances in a river are commonly expressed as biochemical oxygen demand or BOD. BOD is measured as the oxygen that is removed in a 20°C, five-day test and is expressed as BOD_5. This value has be converted to an ultimate BOD or L value at the temperature of interest as:

$$L = \frac{BOD_5}{1 - e^{-5k_{20}}}$$

$$(7-13)$$

where:
k_{20} = BOD decay constant at 20°C.
Typical k_{20} values for organic wastes range from about 0.20 to 0.30.

The rate of oxygen consumption by BOD is related to the ultimate BOD concentration as:

$$r_{BOD} = k_T L \qquad (7-14)$$

Oxygen Deficit

The dissolved oxygen in water is determined by Henry's Law based upon a 20 percent partial pressure in the atmosphere. Saturation values as a function of temperature can be extrapolated from known values of 14.6, 11.3, 9.1, and 7.5 mg/L at 0, 10, 20, and 30°C, respectively.

At a given temperature the dissolved oxygen concentration can be expressed also as a deficit or:

$$D = (C_{sat} - C) \qquad (7-15)$$

where:
C_{sat}, C = saturation and actual dissolved oxygen concentration (mg/L), and
D = dissolved oxygen deficit at given temperature.

Mixing

Assuming complete mixing in a river, the concentration of dissolved oxygen and ultimate BOD in a receiving stream are related as:

$$C_0 = \frac{Q_r C_r + Q_w C_w}{Q_r + Q_w} \qquad (7-16)$$

where:
C_r, C_w = concentration of constituents in river and waste, respectively, and
Q_r, Q_w = flow rate of river and waste, respectively.

Reoxygenation

Reoxygenation occurs by diffusion of oxygen from the atmosphere to the river. The rate of reoxygenation or reaeration is given as:

$$r_{reoxy} = k_2 D \qquad (7-17)$$

where:
k_2 = reaeration coefficient (1/d).

Oxygen Sag Model

The dissolved oxygen concentration is a stream after the addition of oxygen demanding wastes can be expressed in differential form as:

$$\frac{dD}{dt_r} = r_{BOD} - r_{reoxy}$$

(7-18)

where:

t_r = travel time in the river from the point of waste addition (d).

The resulting dissolved oxygen profile is shown in Figure 7-4. After addition of the waste, the dissolved oxygen decreases to a maximum deficit (D_c) or minimum dissolved oxygen level at a distance x_c or travel time t_c. Past this point, the deficit decreases as the river recovers.

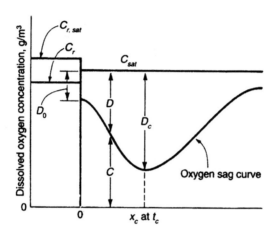

Figure 7-4

Equation 7-18 can be integrated to:

$$D = \frac{kL_0}{k_2 - k}(e^{-kt_r} - e^{-k_2t_r}) + D_0e^{-k_2t_r}$$

(7-19)

where:

D_0, L_0 = oxygen deficit and ultimate BOD concentration at point of waste discharge in the river, respectively.

Water Treatment

Water treatment is used to alter the quality of water to be chemically and bacteriologically safe for human consumption. Common sources are groundwater and surface waters. Groundwater treatment may involve removal of pathogens, removal of iron and manganese, and removal of hardness (Figure 7-5). Surface water treatment typical involves simultaneous removal of pathogens, suspended solids, and taste and odor causing compounds (Figure 7-6). All of these treatment processes are comprised of unit processes.

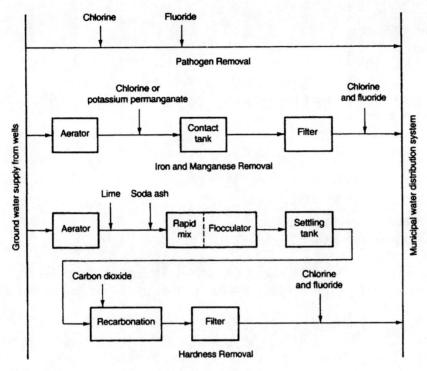

Figure 7-5

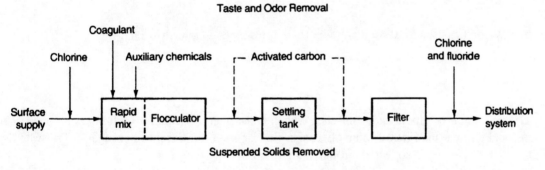

Figure 7-6

Sedimentation

After flocculation, the larger particles are removed by sedimentation. Settling is also used in water treatment after oxidation of iron or manganese. Flocculated particles have densities of about 1400 to 2000 kg/m³.

Settling velocities of particles are determined by Stoke's Law for Reynolds numbers less than 0.3:

$$v_s = \frac{g(\rho_p - \rho_w)d_p^2}{18\mu}$$

(7-20)

where:
v_s = terminal settling velocity (m/s)

g = gravitational constant (9.8 m/s^2);
d_p = particle diameter (m); and
μ = water viscosity (0.001 kg/m-s).

For particles coming into a sedimentation basin, the particles enter at all depths. A critical settling velocity related to the sedimentation basin can be calculated as:

$$v_{sc} = \frac{Q}{A_s}$$

(7-21)

where:
Q = flow rate into sedimentation basin (m^3/s); and
A_s = surface area of sedimentation basin (m^2).

In standard sedimentation theory, if the settling rate of the particles to be removed is greater than v_{sc}, then the particles will be 100% removed. For particles with v_s less than v_{sc}, the particles will only be partially removed with the removal given as:

$$R = \frac{v_s}{v_{sc}}$$

(7-22)

where:
R = decimal removal.

The critical settling velocities used for design are typically expressed as an overflow rate.

Filtration

Filtration in water treatment is commonly accomplished with granular filters. Media for such filters is comprised of sand, charcoal, and garnet; the filters can involves one, two, or several different filter media. The water is applied to the top of the filter and is collected through underdrains.

Filtration is a complex process involving entrapment, straining, and absorption. As the filtration process proceeds, the head loss associated with the water flow through the porous media increases as the filtered material accumulates in the pores. In addition, the number of particles passing through the filter increases with a subsequent increase in effluent turbidity. When either the allowable head loss or effluent quality are exceeded, then the filter has to be backwashed to remove the accumulated material.

Softening
The removal of hardness is accomplished using lime/soda ash softening or ion exchange.

Lime/soda ash softening
Lime/soda ash softening occurs by the addition of calcium hydroxide and sodium bicarbonate to form a chemical precipitate which is removed by sedimentation and filtration.

For waters containing excess, the calcium hydroxide reacts with carbon dioxide and carbonate hardness as:

$CO_2 + Ca(OH)_2 = CaCO_3 + H_2O$

$Ca(HCO_3)_2 + Ca(OH)_2 = CaCO_3 + 2H_2O$

$Mg(HCO_3)_2 + 2Ca(OH)_2 = 2CaCO_3 + Mg(OH)_2 + 2H_2O$

The sodium bicarbonate and calcium hydroxide reacts with the non-carbonate hardness as:

$Ca^{+2} + Na_2CO_3 = CaCO_3 + H_2O$

$Mg^{+2} + Ca(OH)_2 = Mg(OH)_2 + Ca^{+2}$

Recarbonation is required after treatment to remove the excess lime, magnesium hydroxide, and carbonate to reduce the pH to about 8.5 to 9.5.

Based upon the above equations, the requirements of lime (L) and soda ash (SA) in meq/L are:

$$L = CO_2 + HCO_3^- + Mg^{+2} \qquad (7\text{-}23)$$

$$SA = Ca^{+2} + Mg^{+2} - Alk \qquad (7\text{-}24)$$

The lime requirements need to be increased about 1 eq/m^3 to raise the pH to allow precipitation of the magnesium hydroxide.

Ion Exchange

Ion exchange processes are used to remove ions of calcium, magnesium, iron, and ammonium. The exchange material is a solid that has functional groups that replaces ions in solutions for ions on the exchange material. Ion exchange materials can be made to remove cations or anions. Most materials are regenerated by flushing with solutions with high concentrations of either H$^+$ or Na$^+$.

Ion exchange process is controlled by the exchange capacity and the selectivity. The exchange capacity expresses the equivalents of cations or anions that can be exchanged per unit mass. The selectivity for ion B as compared to ion A is expressed as:

$$K_{B/A} = \frac{\chi_{R-A}\left[A^+\right]}{\chi_{R-B}\left[B^+\right]} \qquad (7\text{-}25)$$

where:

$K_{B/A}$ = selectivity coefficient for replacing A on resin with B;

χ_{R-A}, χ_{R-B} = mole fractions of A and B for the absorbed species.

Chlorination

Chlorination is the most common form of disinfection for treatment of water. Chlorine is a strong oxidation agent which destroys organisms by chemical attack. Chlorine is usually added in the form of chlorine gas at a level of 2 to 5 mg/L. Other common forms are chlorine dioxide, sodium hypochlorite, and calcium hypochlorite.

Chlorine in water undergoes an equilibrium reaction to form HOCl and OCl$^-$. The hypochlorite is a weak acid as listed in Table 7-5. HOCl is a more effective disinfectant than OCl$^-$; as such, the effectiveness is strongly dependent upon pH.

7-12

The effectiveness of chlorination depends upon the time of contact, the chlorine concentration, and the concentration of organisms. The effect of time is modeled as:

$$\frac{N_t}{N_0} = e^{-kt^m}$$

(7-26)

where:
N_t, N_0 = number of organisms at time t and time zero;
k = decay coefficient (1/d)
t = time (d)
m = empirical coefficient (usually 1).

Fluoridation
Fluoridation of water supplies has been shown to dramatically reduce dental caries. Optimum concentrations are about 1 mg F/L. The most commonly used compounds are sodium fluoride, sodium silicofluoride, and fluorosilicic acid. Fluoride is usually added after coagulation or lime/soda ash softening since high calcium concentrations can result in precipitation.

Activated Carbon
Activated carbon is added in water treatment to adsorb taste and odor causing compounds and to remove color. The process usually involved the direct addition of powdered activated carbon with removal in the sedimentation and filtration units. The required dosage is usually determined empirically using laboratory tests.

Sludge Treatment
The sources of sludge in water treatment include sand, silt, chemical sludges, and backwash solids. These sludge are typically concentrated in settling basins or lagoons. The sludges can be further concentrated by drying on sand drying beds or by centrifugation.

Problems:

7.1
Find a single pipe to replace the following pipe loop:

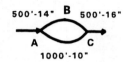

7.2
How many grams of oxygen are required to burn 1 gram of methane?

$$CH_4 + O_2 \rightarrow CO_2 + H_2O$$

7.3
A liter of water is at equilibrium with an atmosphere containing a partial pressure of 0.1 atm of CO_2. How many grams are dissolved in the water? ($\alpha = 2.0$ g/l-atm)

7.4
How much [HOCl] is present in a solution containing 0.1M chlorine at pH 8?

7.5
How many grams of fluoride would be present in a solution saturated with calcium fluoride?

7.6
A 5-d BOD and ultimate BOD are measured at 180 mg/L and 200 mg/L, respectively. What is the decay coefficient?

7.7
A wastewater is discharged to a river with a dissolved oxygen concentration of 1 mg/l. The river is at 20°C and saturated with dissolved oxygen. If the flows are 2.8 x10^{-2} m/s and 2.8 m/s for the waste water and the river, respectively, what is the oxygen deficit after mixing?

7.8
A water treatment plant is required to produce an average treated water flow of 14.4 x10^4 m^3/d from a water source of the following characteristics:

Chlorines:	92 mg/L	Magnesium:	28 mg/L
Potassium:	31 mg/L	Alkalinity:	135 mg/L as CaCO$_3$
Sodium:	14 mg/L	pH:	7.8
Sulfates:	134 mg/L	Temperature	21°C
Calcium:	94 mg/L	Total Solids:	720 mg/L

Determine an ion balance for the water.

7.9

A water contains silt particles with a uniform diameter of 0.02 mm and a specific gravity of 2.6. What removal is expected in a clarifier with an overflow rate of 12 m/d (300 gal/(ft²-d)) ?

Solutions:

7.1

Assume a flow in A-B-C of 4 cfs

h_L in AB = 5.8'/1000'

$\qquad$ = 2.9 ft

h_L in BC = 3.0'/1000'

$\qquad$ = 1.5'

Total h_L = 4.5'

$\qquad$ = 4.5'/1000'

Use ϕ = 14.5", L=1000'

for a h_L of 4.5' for A-C

then Q = 1.4 cfs

Q_{Total} = 4 + 1.4 = 5.4 cfs

Equivalent diameter for loop = 16.2"

Equivalent length for loop = 1000'.

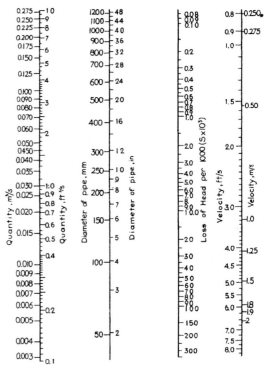

Problem 7.1. Flow in old cast-iron pipes.
(Hazen-Williams C=100).

7.2

Balance the equation as:

$$CH_4 + 2O_2 \rightarrow CO_2 + 2H_2O$$

$$g\ O_2 = 1\ g\ CH_4 \times \frac{1\ mole\ CH_4}{16\ g\ CH_4} \times \frac{2\ moles\ O_2}{1\ mole\ CH_4} \times \frac{32\ g\ O_2}{1\ mole\ O_2}$$

$$= 4\ g\ O_2$$

7.3

$$C_{equil} = \alpha \, P \, gas$$
$$= 2.0 \text{ g / 1- atm} \times 0.1 \text{ atm}$$
$$= 0.2 \text{ g / liter}$$

$$M_{CO_2} = C \times V$$
$$= 0.2 \text{ g / l } \times 1 \text{ liter}$$
$$= 0.2 \text{ g}$$

7.4

$$\left[H^+\right] = 10^{-pH} = 10^{-8}$$

$$HOCl \Leftrightarrow H^+ + OCl^-$$

$$\frac{\left[H^+\right]\left[OCl^-\right]}{\left[HOCl\right]} = 10^{-7.5}$$

$$\frac{\left[H^+\right]\left(0.1 - \left[HOCl\right]\right)}{\left[HOCl\right]} = 10^{-7.5}$$

$$\frac{\left(0.1 - \left[HOCl\right]\right)}{\left[HOCl\right]} = 3.16$$

$$\left[HOCl\right] = 0.024 M$$

7.5

$$CaF_2 \rightarrow Ca^{++} + 2F^-$$

$$\left[Ca^{++}\right]\left[F^-\right]^2 = 3 \times 10^{-11}$$

$$\left[\tfrac{1}{2}F^-\right]\left[F^-\right]^2 = 3 \times 10^{-11}$$

$$\left[F^-\right] = \left[2\left(3 \times 10^{-11}\right)\right]^{\frac{1}{3}}$$

$$= 4.2 \times 10^{-4} M$$

$$= 8.0 \times 10^{-3} \text{ g / liter}$$

7.6

$$1 - e^{-K(5d)} = BOD_5 / L$$
$$= 180 / 300 = 0.6$$
$$K = 0.18 / d$$

7.7

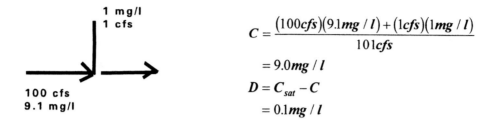

$$C = \frac{(100cfs)(9.1mg \, / \, l) + (1cfs)(1mg \, / \, l)}{101cfs}$$

$$= 9.0mg \, / \, l$$

$$D = C_{sat} - C$$

$$= 0.1mg \, / \, l$$

7.8

(a) The miliequivalents per liter for the ions are:

Ion	Conc(mg/l)	Equiv. Wt.	Conc (meq/l)
Na^+	14	23	0.61
K^+	31	39	0.79
Mg^{++}	28	12.2	2.30
Ca^{++}	94	20	4.7
		Total	8.4
$SO_4^=$	134	48	2.79
Cl^-	92	35.5	2.59
$HCO_3^- + CO_3^=$	135	50	2.70
		Total	8.1

When the concentration of the alkalinity (expressed as $CaCO_3$) is converted to meq/L, it will be equal to the sum of the concentrations of the bicarbonate and the carbonate.

7.9

$$v_s = \frac{g(\rho_p - \rho_w)d_p^2}{18\mu}$$

$$= \frac{\left(9.8 \, m/s^2\right)\left(1600 \, Kg/m^3\right)\left(2 \times 10^{-5}m\right)^2}{(18)\left(1 \times 10^{-3} \, Ns \, / \, m^2\right)}$$

$$= 3.5 \times 10^{-5} \, m \, / \, s$$

$$v_{sc} = \frac{300gal}{ft^2 - d} \times \frac{0.0017m \, / \, hr}{1gal \, / \, ft^2 - d} \times \frac{1hr}{3600s}$$

$$= 1.4 \times 10^{-4} \, m \, / \, s$$

$$v_{sc} = 12m \, / \, d \times \frac{d}{86400s} = 1.4 \times 10^{-4} \, m \, / \, s$$

$$K = \frac{3.5 \times 10^{-5} \, m \, / \, s}{14 \times 10^{-5} \, m \, / \, s} = 0.25 \text{ or } 25\%$$

Chapter 8
Computers and Numerical Methods
Lincoln D. Jones

Introduction

The Fundamentals of Engineering Exam contains seven questions concerning computers in the morning session and three questions in the afternoon session. These questions cover the topics of operating systems, networks, interfaces, spreadsheets, flow charting, and data transmission. Each of the branch specific afternoon exams contain three questions on numerical methods related to that branch.

You should review the glossary of important computer related keywords that is presented at the end of this chapter to ensure you have a basic understanding of this broad general topic. Should you find any term unfamiliar to you, a review of that topic would be warranted. The current exam does not include a programming language such as FORTRAN or BASIC but use of applications such as spreadsheets are included. You should be familiar with one of the popular spreadsheet programs such as Excel, Quattro Pro or 1-2-3.

Number Systems

The number system most familiar to everyone is the decimal system based on the ten symbols 0 through 9. This base 10 system requires ten different digits to create the representation of numbers.

A far simpler system is the binary number system. This base 2 system uses only the characters 0 and 1 to represent any number. A binary representation 110, for example, corresponds to the number

$$1x2^2 + 1x2^1 + 1x2^0 = 4 + 2 + 0 = 6$$

Similarly the binary number 1010 would be

$$1x2^3 + 0x2^2 + 1x2^1 + 0x2^0 = 8 + 0 + 2 + 0 = 10$$

The digital computer is based on the binary system of on/off, yes/no, or 1/0. For this reason it may at times be necessary to convert a decimal number (like 12) to a binary number (for 12 it would be 1100).

Example 1

The binary number 1110 corresponds to what decimal (base 10) number?

Solution

$$1x10^3 + 1x10^2 + 1x10^1 + 0x10^0 = 8 + 4 + 2 = 14$$

Data Storage

Memory chips include random access memory (RAM), read only memory (ROM),and programmable read only memory (PROM).

Diskettes have been the primary way to move data about on personal computers. Initially diskettes were 5-¼ inch "floppies" (360K to 2MB of data) that could be damaged easily. They have since been replaced by 3-½ inch diskettes (720 to several MB of data) in a hard plastic housing.

Hard drives (currently about 500MB to 2GB of data) are units with the hard disks permanently sealed in a module. Other storage devices are removable hard disk cartridges, optical disks, read-only media (ROM), write once, read many media (WORM), and compact disk read-only memory (CD-ROM).

Data Transmission

A computer produces digital signals (represented as on-off or 0-1), but these signals often must be transmitted on voice transmission (analog) systems. The conversion of digital signals to analog signals is called *modulation*. The subsequent conversion back to digital from analog is called *demodulation*. The device doing this conversion is called a *modem* (short for modulation-demodulation). Early modems had speeds of 300-1200 bits per second, but modern modems operate at 14,400-28,800 bits per second.

Data transmission requires that the equipment at both the sending and receiving locations must be able to "talk" to each other. Two methods commonly used are asynchronous and synchronous transmission. In asynchronous transmission a start signal is sent at the beginning and a stop signal at the end of each character. Synchronous transmission, on the other hand, is where a bit pattern is transmitted at the beginning of the message to synchronize the internal clocks of the sending and receiving devices.

Programming

Computer programming may be thought of as a four-step process:
1. Defining the problem
2. Planning the solution
3. Preparing the program
4. Testing and documenting the program

Once the problem has been carefully defined, the basic programming work of planning the computer solution and preparing the detailed program can proceed. In this section the discussion will be limited to two ways to plan a computer program to solve a problem:

Algorithmic Flowcharts
Pseudocode

Algorithmic Flowcharts

An algorithmic flowchart is a pictorial representation of the step-by-step solution of a problem using standard symbols. Some of the commonly used figures are shown in Figure 8-1. Consider the following simple problem.

Example 2

A present sum of money (P) at an annual interest rate (I), if kept in a bank for N years would amount to a future sum (F) at the end of that time according to the equation $F = P(1+I)^N$. Prepare a flowchart for $P = \$100$ $I = 0.07$ $N = 5$ years and compute and output the values of F for all values of N from 1 to 5. Figure 8-2 is a flowchart for this situation.

Figure 8-1 Flowchart Symbols

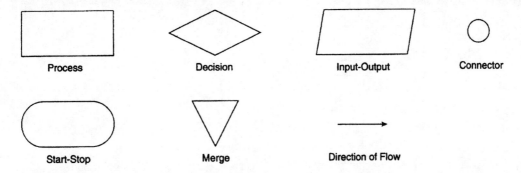

Process

Decision

Input-Output

Connector

Start-Stop

Merge

Direction of Flow

Figure 8-2

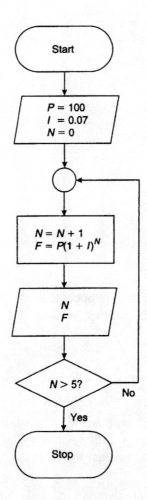

Example 3 ―――――――――――――――――――――――――――――
Consider the flowchart in Figure 8-3.

Figure 8-3

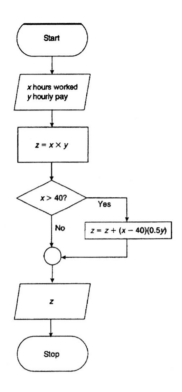

The computation does which of the following?
(a) Inputs hours worked and hourly pay and outputs the weekly paycheck for 40 hours or less.
(b) Inputs hours worked and hourly pay and outputs the weekly paycheck for all hours worked including over 40 hours at premium pay.

The answer is (b).

Pseudocode

Pseudocode is an English-like language representation of computer programming. It is carefully organized to be more precise than a simple statement, but may lack the detailed precision of a flowchart or of the computer program itself.

Example 4 ───

Prepare pseudocode for the computer problem described in example 8-2.

Solution

INPUT P,I, and N = 0
DOWHILE   N < 5
COMPUTE   N = N + 1
          $F = P (1 + I)^N$
OUTPUT   N,F
IF N > 5 THEN ENDO

───

Spreadsheets

For today's engineers the ability to create and use spreadsheets is essential. The most popular spreadsheet programs are Microsoft's **Excel**, Novell's **Quattro Pro**, and Lotus **1-2-3**. Each of these programs use similar construction, methods, operators, and relative references.

Three type of information may be entered into a spreadsheet: text, values, or formulas. Text includes labels, headings and explanatory text. Values are numbers, times or dates. Formulas combine operators and values in an algebraic expression.

A **cell** is the intercept of a column and a row. Its location is based upon its column-row location, for example B3 would be the intercept of column B and row 3. Column labels are across the spreadsheet and row labels are on the side. To change a cell entry, the cell must be highlighted using either an address or pointer.

A group of cells may be called out by using a **range**. Cells A1, A2, A3, A4 could be called out using the range reference A1:A4 (or A1..A4). Similarly, A2,B2,C2,D2 would use the range reference A2:D2 (or A2..D2).

In order to call out a block of cells, a range call out might be A2:C4 (or A2..C4) and would reference the following cells:

    A2 B2 C2
    A3 B3 C3
    A4 B4 C4

Formulas may include cell references, operators (such as +,-,*,/) and functions (such as SUM, AVG). The following formula SUM(A2:A6) or SUM(A2..A6) would be evaluated as equal to A2+A3+A4+A5+A6.

Relational References

Most spreadsheet references are relative to the cell's position. For example if the content of cell A5 contained B4 then the value of A5 is the value of the cell up one and over one. The relational reference is most frequently used in table tabulations as the following example for an inventory where cost times quantity equals value and the sum of the values yields the total inventory cost.

Inventory Valuation

	A	B	C
1 Item	Cost	Quantity	Value
2 box	5.2	2	10.4
3 tie	3.4	3	10.2
4 shoe	2.4	2	4.8
5 hat	1.0	1	1.0
6 Sum			26.4

In C2 the formula A2*B2
In C3 the formula A3*B3 and so forth.
For the Summation use the function SUM. In C6 use SUM(C2:C5)

Instead of typing in each cell's formula, the formula can be copied from the first cell to all of the subsequent cells by first highlighting cell C2 then dragging the mouse to include cell C5. The first active cell is C2 would be displayed in the edit window. Type the formula for C2 as A2*B2 and hold the control key down and the enter key is pressed. This copies the relational formula to each of the highlighted cells. Since the call is relational, then in cell C3 the formula is evaluated as A3*B3. In C4 the cell is evaluated as A4*B4. Similarly, edit operations to include copy, or fill operations simplify the duplication of relational formula from previously filled in cells.

Arithmetic Order of Operation

Operations in equations use the following sequence: exponentiation, multiplication and division, followed by addition and subtraction. Parentheses in formulas override normal operator order.

Absolute References

Sometimes a reference to a cell must be used that should not be changed, such as an data variable. An absolute reference can be named by inserting a dollar sign "$" before the column-row reference. If B2 is the data entry cell, then by using B2 as its reference in another cell, the call will always be evaluated to cell B2, even though it may be copied to another cell. Mixed reference can be made by using the dollar sign for only one of the elements of the reference. The reference B$2 is a mixed reference in that the row does not change but the column remains a relational reference.

The power of spreadsheets makes repetitive calculations very easy. In many operations, periods of time are normally new columns. Each element becomes a row item with changes in time becoming the columns.

All spreadsheet programs allow for changes in the appearance of the spreadsheet. Headings, borders, or type fonts are usual customizing tools.

NUMERICAL METHODS

This portion of numerical methods include techniques of finding roots of polynomials by Routh-Hurwitz criterion and Newton methods, Euler's techniques of numerical integration and the trapezoidal methods, and techniques of numerical solutions of differential equations.

Root Extraction
Routh-Hurwitz Method (Without Actual Numerical Results):

Root extraction, even for simple roots (i.e., without imaginary parts), can become quite tedious. Before attempting to find roots, one should first ascertain whether they are really needed or whether just knowing the area of location of these roots will suffice. If all that is needed is knowing whether the roots are all in the left half plane of the variable (such as in the s-plane when using Laplace transforms -- such as frequently the case in determining system stability in control systems), then one may use the Routh-Hurwitz criterion. This method is fast and easy even for higher ordered equations. As an example, consider the following polynomial:

$$P(x) = \prod_{m=1}^{n}(x-a_m) = x^n + a_1 x^{n-1} + a_2 x^{n-2} + \ldots + a_{n-1} \qquad (8\text{-}1)$$

Here, finding the roots for n>3 can become quite tedious without a computer; however if one only needs to know if any of the roots have positive real parts, then use the Routh-Hurwitz method. Here, an array may be made listing the coefficients of every other term starting with the highest power, n, on a line, then a following line may be made listing the coefficients of the terms left out of the first row. Following rows are constructed using Routh-Hurwitz techniques, and after completion of the array, one merely checks to see if all the signs are the same (unless there is a zero coefficient -- then something else needs to be done) in the first column; if none, no roots will exist in the right half plane. A simple technique used in control systems (for details, see almost any text dealing with stability of control systems. A short example follows:

$F(s) = s^3 + 3s^2 + 2s + 10,$ Array:

	s^3	1	2
	s^2	3	10
	s^1	-4/3	0
	s^0	10	0

Where the s^1 term is formed as: $(3 \times 2 - 10 \times 1)/3 = -4/3$. For details refer to any text on control systems or numerical methods.

$= (s+?)(s+?)(s+?)$

Here, there are two sign changes, one from 3 to -4/3, and one from -4/3 to 10; this means there will be two roots in the right half plane of the s-plane, which yields an unstable system; this technique represents a great savings in one's time without having to actually factor the equation.

Newton's Method
The use of Newton's method of solving a polynomial and the use of iterative methods can greatly simplify the problem. This method utilizes synthetic division and is based upon the remainder theorem. This synthetic division requires estimating a root at the start, and, of course, the best estimate is the actual root. The root is the correct one when the remainder is zero. (There are several ways of estimating this root, including a slight modification of the Routh-Hurwitz criterion.)

By taking a $P_n(x)$ polynomial (see equation 8-1) and dividing it by an estimated factor $(x-x_1)$, the result is a reduced polynomial of degree n-1, $Q_{n-1}(x)$, plus a constant remainder of b_{n+1}. Thus, another way of describing equation 8-1 is,

$$P_n(x)/(x-x_1) = Q_{n-1}(x) + b_{n-1} / (x-x_1) \quad \text{or as} \quad P_n(x) = (x-x_1)Q_{n-1}(x) + b_{n-1} \tag{8-2}$$

If one lets $x=x_1$, the equation 8-2 becomes,

$$P_n(x=x_1) = (0)Q_{n+1}(x) + b_{n+1} = b_{n+1} . \tag{8-3}$$

Equation 8-3 leads directly to the remainder theorem: "The remainder on division by $(x-x_1)$ is the value of the polynomial at $x= x_1$, $P_n(x_1)$." *

    Newton's method (actually, the Newton-Raphson method) for finding the roots for an nth order polynomial is an iterative process involving obtaining an estimated value of a root (leading to a simple computer program). The key to the process is getting the first estimate of a possible root; without getting too involved, recall the coefficient of x^{n-1} represents the sum of all of the roots, and the last term represents the product of all n roots, then the first estimate can be "guessed" within a reasonable magnitude. After a first root is chosen, find the rate of change of the polynomial at the chosen value of the root to get the next closer value of the root, x_{n+1}. Thus the new root estimate is based on the last value chosen,

$$\boxed{x_{n+1} = x_n - P(x_n)/P'(x_n), \quad \text{where } P'(x_n) = dP(x)/dx \text{ evaluated at } x=x_n} \tag{8-4}$$

NUMERICAL INTEGRATION

Numerical integration routines are extremely useful in almost all simulation type programs, design of digital filters, theory of z-transforms, and almost any problem solution involving differential equations. And since digital computers have essentially replaced analog computers (which were almost true integration devices), the techniques of approximating integration are well developed. Several of the techniques are briefly reviewed below.

Euler's Method

For a simple first order differential equation, say $dx/dt + ax = af$, one could write the solution as a continuous integral or as an interval type one:

$$x(t) = \int^{t} [-ax(\tau)+af(\tau)]d\tau \tag{8-5a}$$

$$x(kT) = \int^{kT-T} [-ax+af]d\tau + \int_{kT-T}^{kT} [-ax+af]d\tau = x(kT-T) + A_{rect} \tag{8-5b}$$

Here, A_{rect} is the area of $(-ax+af)$ over the interval $(kT-T) < \tau < kT$. One now has a choice looking back over the rectangular area or looking forward. The rectangular width is, of course, T. For the forward looking case presented in a first approximation for x_1 is**,

$$x_1(kT) = x_1(kT-T) + T[ax1(kT-T)+af(kT-T)]T = (1-aT)x_1(kT-T) + aTf(kT-T) \tag{8-5c}$$

---

*Gerald & Wheatley, *Applied Numerical Analysis*, Addison-Wesley, 3rd ed., 1985.
**This method is as presented in Franklin & Powell, *Digital Control of Dynamic Systems*, Addison-Wesley, 1980, page 55.

Or, in general, for Euler's forward rectangle method, the integral may be approximated in its simplest form (using the notation $t_{K+1} - t_K$ for the width, instead of T which is kT-T) as,

$$\int_{t_k}^{t_{k+1}} x(\tau)d\tau \approx (t_{k+1} - t_k)x(t_k) \qquad (8\text{-}6)$$

Trapezoidal Rule

This trapezoidal rule is based upon a straight line approximation between a function, f(t), at t_0 and t_1. To find the area under the function, say a curve, is to evaluate the integral of the function between point a and b. The interval between these points are subdivided into subintervals; the area of each subinterval is approximated by a trapezoid between the end points. It will only be necessary to sum these individual trapezoids to get the whole area; by making the intervals all the same size, the solution will be simpler. For each interval of delta t (i.e., $t_{K+1} - t_K$), the area is then given by,

$$\boxed{\int_{t_K}^{t_{k+1}} x(\tau)d\tau \approx (1/2)(t_{k+1} - t_k)[x(t_{k+1}) + x(t_k)]} \qquad (8\text{-}7)$$

This equation gives good results if the delta t's are small but it is for only one interval and is called the "local error". This error may be shown to be $-(1/12)(\text{delta } t)^3 f''(t=\xi_1)$, where ξ_1 is between t_0 and t_1. For a larger "global error" it may be shown that,

$$\text{Global error} = -(1/12)(\text{delta } t)^3 [f''(\xi_1) + f''(\xi_2) + \ldots + f''(\xi_n)] \qquad (8\text{-}8)$$

Following through on equation 8-8 allows one to predict the error for the trapezoidal integration; This technique is beyond the scope of this review or probably the examination; however for those interested, please refer to pages 249-250 of the previous mentioned reference to Gerald & Wheatley.

NUMERICAL SOLUTIONS OF DIFFERENTIAL EQUATIONS

This solution will be based upon a first order ordinary differential equations. However the method may be extended to higher ordered equations by converting them to a matrix of first ordered ones.

Integration routines produce values of system variables at specific points in time and update this information at each interval of delta time as T (delta $t = T = t_{k+1} - t$). Instead of a continuous function of time, $x(t)$, the variable x will be represented with discrete values such that $x(t)$ is represented by x_0, x_1, x_2, ..., x_n. Consider a simple differential equation as before, as (based upon Euler's method),

$$dx/dt + ax = f(t).$$

Now assume the deta time periods, T, are fixed (not all routines use fixed step sizes), then one writes the continuous equation as a difference equation where: $dx/dt \approx (x_{k+1} - x_k) / T = -ax_k + f_k$ or, solving for the updated value, x_{k+1},

$$(x_{k+1}) = x_k - Tax_k + Tf_k \quad \text{for fixed increments.} \tag{8-9a}$$

By knowing the first value of $x_{k=o}$ (or the initial condition), the solution may be achieved for as many "next values" of x_{k+1} as desired for some value of T. The difference equation may be programmed in almost any high level language on a digital computer; however, T must be small as compared to the shortest time constant of the equation (here, $1/a$).

The following equation (with the "f" term meaning a "function of" rather than as a "forcing function" term -- as used in equation 8-5a) is in a more general form of equation 8-9a. This equation is obtained by letting the notation (x_{k+1}) become $y[k+1)\Delta t]$ and is written (perhaps somewhat more confusing) as,

$$\boxed{y[(k+1)\Delta t] \ y(k\Delta t) + \Delta t f[(y(k\Delta t), k \Delta t]]} \tag{8-9b}$$

Reduction of Differential Equation Order

To reduce the order of a linear time dependent differential equation, the following technique is used. For example, assume a second order one:

$$x'' + ax' + bx = f(t), \text{ define } x = x_1 \text{ and } x' = x_1' = x_2, \text{ then, } x_2' + ax_2 + bx_1 = f(t),$$

$$x_1' = x_2 \qquad \text{<--- By definition.}$$
$$x_2' = -bx_1 + ax_2 + f(t)$$

This equation can be extended to higher order systems and, of course, be put in a matrix form (called the state variable form). And it can easily be set up as a matrix of first order difference equation for solving digitally.

Packaged Programs

Most currently available packaged simulation programs use algorithms not necessarily based upon Euler's methods but more advanced methods such as the Runge- Kutta method. Automatic variable step size methods like Milne's may also be used. However, as mentioned before, these routines are all built into the packaged programs and may be transparent to the user. The user of a specialized program may be without knowledge of the high level language being employed (except for certain modifications).

Glossary of Computer Terms

Accumulators	Registers which hold data, address, or instructions for further manipulations in the ALU
Address bus	Two way parallel path connecting processors and memory containing addresses
AI	Artificial Intelligence
Algorithm	A sequence of steps applied to a given data set which solve the intended problem
Alphanumeric data	Data containing the characters a,b,c...z,0,1,2,...9

ALU	Arithmetic and logic unit
ASCII	American Standard Code for Information Interchange ,7 bit/character (Pronounced AS-key)
Asynchronous	Form of communications which message data transfer is not synchronous with the basic transfer rate requiring start/stop protocol
Baud rate	Bits per second
BIOS	Basic input/output system
Bit	0 or 1
Buffer	Temporary storage device
Byte	8 bits
Cache Memory	Fast look-ahead memory connecting processors with memory offering faster access to often used data
Channel	Logic path for signals or data
CISC	Complex instruction-set control
Clock rate	cycles per second
Control bus	Separate physical path for control and status information
Control unit	Fetches, decodes instructions to control the operations of registers and ALU's
CPU	Central Processing Unit, primary processor
Data buffer	Temporary storage of data
Data bus	Separate physical path dedicated for data
Digital	Discrete level or valued quantification vs analog or continuous valued
Duplex Communication	Communications mode where data is transmitted in both directions, but only in one direction at any one time
Dynamic memory	Storage which must be continually hardware refreshed to remain
EBCDIC	Extended Binary Coded Decimal Interchange Code - 8 bits/character (Pronounced EB-see-dick)
EPROM	Erasable programmable read-only memory
Expert systems	Programs with AI which learn rules from external stimuli
Floppy disk	Removable disk media in various sizes, 5-¼", 3-½"
Flowchart	Graphical depiction of logic using shapes and lines
G-byte	Giga bytes 1,073,741,824 bytes or 2^{30}
Half-duplex communication	2-way communications path in which only 1 direction operates at a time (transmit or receive)
Handshaking	Communications protocol to start/stop data transfer
Hard disk	Disk which has non-removable media
Hardware	Physical elements of a system
Hexadecimal	Numbering system (base 16) uses 0-9,A,B,...F
Hierarchical database	Database organization containing hierarchy of indexes/keys to records
I/O	Input/Output devices such as terminal, keyboard, mouse, printer

IR	Instruction register
K-bytes	Kilo-bytes 1024 bytes or 2^{10}
LAN	Local Area Network
LIFO	Last In-First Out
LSI	Large scale integration
Main memory	That memory seen by CPU
M-bytes	Mega-bytes 1,048,5 76 bytes or 2^{20}
Memory	Generic term for random access storage
Microprocessor	Computer architecture with Central Processing Unit in one LSI chip
MODEM	Modulator-demodulator
MOS	Metal Oxide Semiconductor
Multiplexer	Device which switches several input sources one at a time to an output
Narrow band	Frequency domain reference to channel band width versus baseband
Nibbles	4 bits
Non-volatile memory	As opposed to volatile memory, does not need power to retain present state
Number systems	Method of representing computer data into human readable information
OCR	Optical Character Recognition
OS	Operating system
OS memory	Memory dedicated to the OS, not useable for other functions
Parallel interface	A character(8 bit) or word(16 bit) interface with as many wires as bits in interface plus data clock wire.
Parity	Method for detecting errors in data, 1 extra bit carried with data, even or odd, to make the sum of one bits in a data stream
PC	Program counter, or personal computer
Peripheral devices	Input/Output devices not contained in Main processing hardware
Program	A sequence of computer instructions
PROM	Programmable Read-Only Memory
Protocols	Established set of handshaking rules enabling communications
Pseudocode	An English-like way of representing structured programming control structures
RAM	Random Access Memory
Real time/Batch	Method of program execution; Real-time implies immediate execution, Batch mode is postponed until run on a group of related activities
Relational database	Database organization which related individual elements to each other without fixed hierarchical relationships
RISC	Reduced instruction set computer
ROM	Read Only Memory
Scratchpad memory	Highspeed memory either in hardware or software
Sequential Storage	Memory (usually tape) accessed only in sequential order (n, n + 1,..)

Serial Interface	Single data stream which encodes data by individual bit per clock period
Simplex communication	One-way communications
Software	Programmable logic
Stacks	Hardware memory organization implementing Last In-Last Out access
Static memory	Memory which does not require intermediate refresh cycles to retain state
Structured Programming	Use of Structured programming constructs such as Do-While, If-Then, Else
Synchronous	Communications mode which data and clock are at same rate
Transmission speed	Rate at which data is moved in baud. (bits per second,bps)
Virtual memory	Addressible memory outside physical address bus limits through use of memory mapped pages
Volatile memory	Memory whose contints are lost when power is removed.
VRAM	Video memory
Wide-band	300-3300 Hz
Words	8,16,or 32 bits
WYSIWYG	What you see is what you get
16-bit	Basic organization of data with 2-bytes per word
32-bit	Basic organization of data with 4-bytes per word
64-bit	Basic organization of data with 8-bytes per word
80386	Intel's microprocessor architecture based on 16 data address bus, extended virtual memory, external math coprocessor
80486	Upgrade to 80386 incorporating Math coprocessor within VLSI
80586	Intels microprocessor architecture based on 16 bit data, 32 bit address bus

8-14

Problems and Solutions

8-1. In spreadsheets, what is the easier way to write
B1 + B2 + B3 + B4 + B5
(a) Sum(B1:B5)
(b) (B1..B5)Sum
(c) @B1..B5SUM
(d) @SUMB2..B5

8-2. The address of the cell located at column 23 and row C is
(a) 23C
(b) C23
(c) C.23
(d) 23.C

8-3 Which of the following is not correct?

(a) A CD-ROM may not be written to by a PC.
(b) Data stored in a batch processing mode is always up-to-date.
(c) The time needed to access data on a disk drive is the sum of the seek time, the head switch time, the rotational delay, and the data transfer time.
(d) The methods for storing files of data in secondary storage are: sequential file organization, direct file organization, and indexed file organization.

8-4. Which of the following is false?
(a) Flowcharts use symbols to represent input/output, decision branches, process statements and other operations.
(b) Pseudocode is an English-like description of a program.
(c) Pseudocode used symbols to represent steps in a program.
(d) Structured programming breaks a program into logical steps or calls to subprograms.

8-5. In Pseudocode using DOWHILE , the following is true:
(a) DOWHILE is normally used for decision branching.
(b) The DOWHILE test condition must be false to continue the loop.
(c) The DOWHILE test condition tests at the beginning of the loop.
(d) The DOWHILE test condition tests at the end of the loop.

8-6. A Spreadsheet contains the following formulas in the cells:

	A	B	C
1		A1+1	B1+1
2	A1^2	B1^2	C1^2
3	Sum(A1:A2)	Sum(B1:B2	Sum(C1:C2)

If 2 is placed in cell A1, what is the value in Cell C3?
   (a) 12
   (b) 20
   (c)   8
   (d) 28

8-7. A matrix contains the following:

	A	B	C	D
1		3	4	5
2	2	A$2		
3	4			
4	6			

If you copy the formulas from B2 into D4, what is the equivalent formula in D4?
   (a) A$2
   (b) C4
   (c) C4
   (d) C$2

8-8. A Processing system is processor limiter when performing sorting of small tables in memory. Which would speed up computations?
   (a) Adding more main memory
   (b) Adding virtual memory
   (c) Adding cache memory
   (d) Adding peripheral memory

8-9. A small PC processing system is performing large (1MB) matrix operations which is currently I/O limited since memory is limited to 1Mbyte. Which would speed up computations?
   (a) Adding more main memory
   (b) Adding virtual memory
   (c) Adding cache memory
   (d) Adding peripheral memory

8-10. Transmission Protocol: Serial, asynchronous, 8 bit ASCII, 1 Start, 1 Stop, 1 Parity bit, 9600 bps. How long will it take for a 1Kbyte file to be transmitted through the link?
   (a) 0.85 sec.
   (b) 0.96 sec.
   (c) 1.07 sec.
   (d) 1.17 sec.

8-11. Transmission Protocol: Serial, synchronous, 8 bit ASCII, 1 Parity bit, 9600 bps. How long will it take for a 1Kbyte file to be transmitted through the link?
    (a) 0.85 sec.
    (b) 0.96 sec.
    (c) 1.07 sec.
    (d) 1.17 sec.

8-12. The binary representation 10101 corresponds to which base 10 number?
    (A) 3
    (B) 16
    (C) 21
    (D) 10,101

8-13. For the base 10 number of 30, the equivalent binary number is
    (A) 1110
    (B) 0111
    (C) 1111
    (D) 11110

8-14. On personal computers the data storage device with the largest storage capacity is most likely to be
    (A) 3½" diskette
    (B) hard disk
    (C) random access memory
    (D) 3½" HD diskette

8-15.

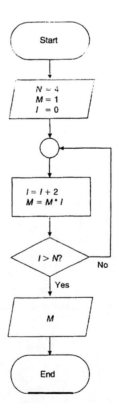

Figure 8-4.

The output value of M in Figure 8-4 is closest to
- (A) 1
- (B) 2
- (C) 8
- (D) 48

8-16. Pseudocode can best be described as
- (A) A simple letter-substitution method of encryption.
- (B) An English-like language representation of computer programming.
- (C) A relational operator in a database.
- (D) The way data are stored on a diskette.

8-17.

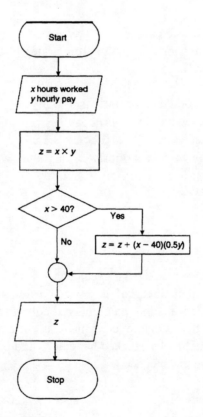

Figure 8-5

8-18

The pseudocode that best represents Figure 8-5 is:

    (A)   INPUT X,Y
            WAGE = X x Y
            IF X less than 40 THEN
            WAGE = WAGE + OVERTIME PAY
            END IF
            OUTPUT WAGE

    (B)   INPUT hours worked and hourly pay
            WAGE = hours worked x hourly pay
            IF hours worked greater than 40 THEN
            WAGE = WAGE + (hours worked -40)(overtime wage supplement)
            END IF
            OUTPUT WAGE

    (C)   INPUT hours worked and hourly pay
            WAGE = (hours worked -40)(overtime rate) + 40(hourly pay)
            OUTPUT WAGE

    (D)   PRINT hours worked and hourly pay
            WAGE = hours worked x hourly pay
            IF hours worked equals 40 THEN
            WAGE = hours worked x hourly pay + (hours worked -40) x overtime pay
            END IF
            OUTPUT WAGE

8-18.

All of the following is true of spreadsheets except:
    (A)   A cell may contain label, value, formula or function
    (B)   A cell reference is made by a numbered column and lettered row reference
    (C)   A cell content that is displayed is the result of a formula entered in that cell
    (D)   Line graphs, bar graphs, stacked bar graphs and pie charts are typical graphs created from spreadsheets.

Solutions

SOLN 8- 1
Sum(B1:B5) or @Sum(B1..B5) Answer is (a).

SOLN 8-2.
Answer is (b).

SOLN 8-3.
Batch mode processing always have delays in updating the database. Answer is (b).

SOLN 8-4.
Pseudocode does not use symbols but uses English-like statements such as IF-THEN, and DOWHILE. Answer is (c).

SOLN 8-5.
IFTHEN is normally used for branching. The DOWHILE test condition must be true to continue branching and the test is done at the beginning of the loop. The DOUNTIL test is done at the end of the loop. Answer is (c).

SOLN 8-6.
Plugging 2 into spreadsheet produces the following matrix:

	A	B	C
1	2	3	4
2	4	9	16
3	6	12	20

The value of C# is 20. Answer is (b).

SOLN 8-7.
The formula contains mixed references. The "$" implies absolute row reference while the column is relative. The result of any copy would eliminate any answer except for absolute row 2 entry relative column reference "A" gets replaced by "C". The cell contains C$2. The answer is (d).

SOLN 8-8.
Processor limited sorting on small tables suggest either speeding up processor cycles or providing faster memory. Since speeding up clock is not an option, then making memory faster is the answer. Adding more memory, or virtual memory does nothing for small tables. Only adding cache memory would allow the CPU to fetch recently used data without full memory cycles thereby speeding up the sorting process. Answer is (c).

SOLN 8-9.
Processing is I/O limited since all of matrix can not fit into memory. Since this is a large matrix, cache memory probably would not effect processing. The best solution is adding more main memory to fit this problem entirly in memory. Answer is (a).

SOLN 8-10.
> 1 Kbyte=2^{10} bytes=1024
> 1024 bytes + 3 overhead bits/byte = 1024(8 bits/byte +3 bits/byte overhead)
> $= 11,264$ bits
> Minimum transmission time = 11264 bits/9600 = 1.17 sec.   Answer is (d).

SOLN 8-11.
> 1 Kbyte=2^{10} bytes=1024
> 1024 bytes + 1 overhead bit/byte = 1024(8 bits/byte + 1 bit/byte overhead)
> $= 9216$ bits
> Minimum transmission time = 9216 bits/9600 = 0.96 sec. Answer is (b).

SOLN-8-12
The binary (base 2) number system representation is

	2^4	2^3	2^2	2^1	2^0	
	16	8	4	2	1	
Binary number	1	0	1	0	1	
	16	0	4	0	1	= 21

Answer is (C).

SOLN-8-13.

2^5	2^4	2^3	2^2	2^1	2^0	
32	16	8	4	2	1	
1	1	1	1	0	= 30	

Answer is (D).

SOLN-8-14. Answer is (B).

SOLN-8-15. Answer is (D).

SOLN-8-16. Answer is (B).

SOLN 8-17. Answer is (B).

SOLN-8-18.
A cell reference is made by a lettered column and a numbered row, example C3.
Answer is (B).

Chapter 9
Legal and Professional Aspects
ETHICS

Engineers, as members of a profession, are expected to conduct themselves in an ethical manner. By this is meant that engineers must be aware of the standards of professional conduct (the ethical standards) and abide by them.

There are codes of ethics that have been prepared by the various national engineering societies, the engineering accreditation board (ABET), and by the National Council of Examiners for Engineering and Surveying (NCEES). The one reproduced in the Fundamentals of Engineering Reference Manual is the *NCEES Model Rules of Professional Conduct*. Thus it is reasonable to assume that it is this code of ethics on which the exam is based.

NCEES Model Rules Of Professional Conduct*
PREAMBLE

To comply with the purpose of the (identify state, registration statute) which is to safeguard life, health, and property, to promote the public welfare, and to maintain a high standard of integrity and practice, the (identify state board, registration statute) has developed the following "Rules of Professional Conduct." These rules shall be binding on every person holding a certificate of registration to offer or perform engineering or land surveying services in this state. All persons registered under (identify state registration statute) are required to be familiar with the registration statute and these rules. The "Rules of Professional Conduct" delineate specific obligations the registrant must meet. In addition, each registrant is charged with the responsibility of adhering to standards of highest ethical and moral conduct in all aspects of the practice of professional engineering and land surveying.

The practice of professional engineering and land surveying is a privilege, as opposed to a right. All registrants shall exercise their privilege of practicing by performing services only in the areas of their competence according to current standards of technical competence.

Registrants shall recognize their responsibility to the public and shall represent themselves before the public only in an objective and truthful manner.

They shall avoid conflicts of interest and faithfully serve the legitimate interests of their employers, clients, and customers within the limits defined by these rules. Their professional reputation shall be built on the merit of their services and they shall not compete unfairly with others.

The "Rules of Professional Conduct" as promulgated herein are enforced under the powers and vested by (identify state enforcing agency). In these rules, the word "registrant" shall mean any person holding a license or a certificate issued by (identify state registration agency).

* Reproduced by permission of NCEES, the copyright owner.

I. REGISTRANT'S OBLIGATION TO SOCIETY

a. Registrants, in the performance of their services for clients, employers, and customers, shall be cognizant that their first and foremost responsibility is to the public welfare.

b. Registrants shall approve and seal only those documents and surveys that conform to accepted engineering and land surveying standards and safeguard the life, health, property, and welfare of the public.

c. Registrants shall notify their employer or client and such other authority as may be appropriate when their professional judgment is overruled under circumstances where the life, health, property, or welfare of the public is endangered.

d. Registrants shall be objective and truthful in professional reports, statements, or testimony. They shall include all relevant and pertinent information in such reports, statements, or testimony.

e. Registrants shall express a professional opinion publicly only when it is founded upon an adequate knowledge of the facts and a competent evaluation of the subject matter.

f. Registrants shall issue no statements, criticisms, or arguments on technical matters which are inspired or paid for by interested parties, unless they explicitly identify the interested parties on whose behalf they are speaking, and reveal any interest they have in the matters.

g. Registrants shall not permit the use of their name or firm name by, nor associate in the business ventures with, any person or firm which is engaging in fraudulent or dishonest business or professional practices.

h. Registrants having knowledge of possible violations of any of these "Rules of Professional Conduct" shall provide the state board information and assistance necessary to the final determination of such violation.

II. REGISTRANTS OBLIGATION TO EMPLOYER AND CLIENTS

a. Registrants shall undertake assignments only when qualified by education or experience in the specific technical fields of engineering or land surveying involved.

b. Registrants shall not affix their signatures or seals to any plans or documents dealing with subject matter in which they lack competence, nor to any such plan or document not prepared under their direct control and personal supervision.

c. Registrants may accept assignments for coordination of an entire project, provided that each design segment is signed and sealed by the registrant responsible for preparation of that design segment.

d. Registrants shall not reveal facts, data, or information obtained in a professional capacity without the prior consent of the client or employer as authorized or required by law.

e. Registrants shall not solicit or accept financial or other valuable consideration, directly or indirectly, from contractors, their agents, or other parties in connection with work for employers or clients.

f. Registrants shall make full prior disclosures to their employers or clients of potential conflicts of interest or other circumstances which could influence or appear to influence their judgment or the quality of their service.

g. Registrants shall not accept compensation, financial or otherwise, from more than one party for services pertaining to the same project, unless the circumstances are fully disclosed and agreed to by all interested parties.

h. Registrants shall not solicit or accept a professional contract from a governmental body on which a principal or officer of their organization serves as a member. Conversely, registrants serving as members, advisors, or employees of a governmental body or department, who are the principals or employees of a private concern, shall not participate in decisions with respect to professional services offered or provided by said concern to the governmental body which they serve.

III. *REGISTRANTS OBLIGATION TO OTHER REGISTRANTS*

a. Registrants shall not falsify or permit misrepresentation of their, or their associates', academic or professional qualifications. They shall not misrepresent or exaggerate their degree of responsibility in prior assignments nor the complexity of said assignments. Presentations incident to the solicitation of employment or business shall not misrepresent pertinent facts concerning employers, employees, associates, joint ventures or past accomplishments.

b. Registrants shall not offer, give, solicit, or receive, either directly or indirectly, any commission or gift, or other valuable consideration in order to secure work, and shall not make any political contribution with the intent to influence the award of a contract by public authority.

c. Registrants shall not attempt to injure, maliciously or falsely, directly or indirectly, the professional reputation, prospects, practice or employment of other registrants, nor indiscriminately criticize other registrants' work.

Example 1——————————————————————————————
The *NCEES Model Rules of Professional Conduct* allows an engineer to do which one of the following?
(a) Accept money from contractors in connection with work for an employer or client.
(b) Compete with other engineers in seeking to provide professional services.
(c) Accept a professional contract from a governmental body even though a principal or officer of the engineer's firm serves as a member of the governmental body.

(d) As the coordinator of an entire project, an engineer may sign or seal all design segments of the project.

Solution
Although the other items are not allowed by the *Model Rules*, nowhere does it say that an engineer cannot compete with other engineers in seeking to provide professional services. But, of course, he/she should do it in an ethical manner.
Answer is (b).

No code of ethics can cover all the kinds of situations an engineer will encounter. It does, however describe in broad terms the ethical principles by which an engineer should be guided.

The examination asks ethics questions concerning relations with clients, peers, and the public. The *NCEES Model Rules of Professional Conduct* are broken down in these same three categories. Thus the exam ethics questions likely are written to see if the examinee has read and generally understands the standards of professional conduct.

Legal and Contract Terms

Terms

(a) How many parties constitute the minimum number needed to form a contract?
(b) What are the essential ingredients of a contract?
(c) What is the difference between a "formal" and an "informal" contract?
(d) What is meant by common law?
(e) What is meant by statute law?
(f) What is equity?
(g) Define the terms "owner", "engineer", and "contractor" as they may appear in a contract document.
(h) What is the legal importance attached to contract plans and specifications.
(I) When construction is performed by "force account" what does this mean?
(j) What is meant by the term "incorporate by reference?"
(k) Why does one specify "or equal" or "an approved equivalent" in a specification for materials? Who does the approving?
(1) Distinguish between the terms "liquidated damages" and "punitive damages."
(m) Describe a surety bond. What is its purpose?

Definitions
(a) Contracts may be formed by two or more parties. That is, there must be a party to make an offer and a party to accept.
(b) To be enforceable at law, a contract must contain the following essential elements:
    1.   There must be a Real agreement.
    2.   The subject matter must be lawful.
    3.   There must be a valid consideration.
    4.   The parties must be legally competent.
    5.   The contract must comply with the provisions of the law with regard to form.

(c) A formal contract depends upon a particular form, or mode of expression, for legal efficacy. All other contracts are called informal contracts since they do not depend upon mere formality for their legal existence.

(d) Common law comprises the body of those rules of action and principles which derive their authority solely from usage and customs.

(e) Statute law consists of acts or statutes established by legislative action.

(f) Equity is a system of doctrines supplementing common and statute law, I.-e., the Maxims of Equity.

(g) The "owner" as specified is the principal. The "engineer" can be a party to an agency with the principal, or he can be an independent contractor. Control and responsibility must be clearly defined. The "contractor" may also be a party to an agency or an independent contractor.

(h) The specifications are written instructions which accompany and supplement the plans. Whereas the drawings show the physical characteristics of the work, the specifications cover the quality of the materials, workmanship, and other technical requirements. The plans and specifications form the guide and standards of performance which will be required.

(I) When the owner elects to do work with his own forces, instead of employing a construction contractor, this is known as the force account method. Under this method, the owner maintains direct supervision over the work, furnishes all materials and equipment, and employs workmen on his own payroll.

(j) The term "incorporate by reference" is when a document is made legally binding by being referenced within a contract, although it is not attached to or reproduced in the contract. This is necessary to eliminate unnecessary repetition.

(k) The terms "or equal" or "an approved equivalent" are used in specifications for materials for several reasons. They take care of the problem of replacement if the original is not available. The terms also allow the bidder to bid lower if he has a supplier with different but equal material at lower cost. The Engineer is the party to do the approving.

(l) Liquidated damages applies when a specific sum of money has been expressly stipulated as the amount of damages to be recovered by either party for a breach of the agreement by the other. Punitive damages are used to punish the defendant in certain situations involving willful, wanton, malicious, or negligent torts.

(m) Surety bonds are normally used in connection with competitive-bid contracts, namely, bid bonds, performance bonds, and payment bonds. The bonds are issued by a third party to guarantee the faithful performance of the contractor.

Problems and Solutions

9-1. Jack, a consulting environmental engineer, was hired by the local garbage company to represent them in hearing before the local town council. In preparing for an appearance before the town council, Jack worked at the garbage company truck storage yard. As he was starting to leave late one evening he discovered that a garbage company employee was moving garbage trucks to the edge of a stream embankment and dumping the liquid that came from the garbage into the stream. The next morning Jack spoke to the truck storage yard manager about what he had seen. The manager told Jack that they had always done it that way and for him to forget it.

Jack feels very uncomfortable with the situation for he knows the dumping is a clear violation of an environmental regulation. Consider Jack's four alternatives and select the best one.

(a) Terminate his consulting with the garbage company.

(b) Assume that since it has gone on for a long time there is no major hazard being created

and forget what he saw.
(c) At the next opportunity advise the local town council of the situation.
(d) Write a letter to the garbage company president outlining what he discovered and advising the president that the dumping practice is an environmental regulation violation.

9-2. Dave's employer is a small engineering firm. For one of his first jobs as a young engineer Dave was given the job of writing the specifications and purchasing a ten-inch pump. He wrote the specifications and sent a Sealed Bid Request to six firms. Only five firms submitted the bids before the bid request deadline. One the last day Dave called the sixth firm to remind them of the sealed bid request. The estimator indicated he was very busy and did not want to take the time to prepare a sealed bid unless he had a good chance of supplying the pump. Dave wants the sixth quote, and wonders what to do.

(a) Since he doesn't plan to open the sealed bids right away, Dave offers to let this last firm submit its quote a week late.
(b) Dave opened the five sealed bids and read them over the phone to the estimator. He said he thought he could beat them and submit a bid by the sealed bid deadline later that day.
(c) Dave already had five bids so he called the sixth firm back and told them there was nothing he could do for them.
(d) Dave opened the five bids and found the lowest one was $8750. He called the estimator for the sixth firm and said, "I can't tell you the bids, but unless you can get under about $8800 forget about it."

9-3. John, a consulting soils engineer, submitted a draft of his report on soil conditions at a plant site to his client's chief engineer. The chief engineer read the draft of the report and indicated to John that he wanted several portions of the report changed so the planned building could be designed with a smaller, less expensive foundation. The chief engineer told John the allowable soil pressure value he wanted him to recommend in the final report.
    Following this meeting John went back to his own office to decide what to do. He believes he has four alternatives. Which one should he select?
(a) Submit the draft soil report as the final report without any changes.
(b) Reexamine his field test data to see if the chief engineer's desired allowable soil pressure value could be recommended. If so, he would do it. If not, he would leave the report as it is.
(c) Make the changes requested by the clients chief engineer and submit the final report.
(d) Write to the president of his client's firm and describe the request of the chief engineer. Advise the president that the draft report is being submitted as the final report.

9-4. Jim, a newly hired engineer, was asked by his supervisor to see what he could learn about their competitors future product plans. This is an important assignment and Jim is anxious to do a good job. He has devised four alternatives. Which one should he select?

(a) Jim could call the competitors and tell them he is a college student working on a report for one of his classes. Jim thinks the competitors would be very forthright in this situation.
(b) Jim could call up some college buddies who are now working for the competitors. A

lunch together could easily result in trading some confidential product plans if Jim worked at it.

(c) Jim could call the competitors and simply ask about their future product plans. If the competitors asked who Jim is, he will honestly tell them who his employer is. But if they do not ask, then Jim will not volunteer the information.

(d) Jim will contact the competitors and at the beginning of the conversation explain who he is, his employer, and that he has been assigned to study competitors future product plans.

9-5. When Glenn, a product development engineer, discovered that his manufacturing company was about to lose a major customer to an overseas competitor, he resolved to warn the customer that he might be making a serious mistake. He is considering which one of the following statements to make to the customer:

(a) I read recently that a lot of this kind of product is being assembled in the foreign country by prison labor. I hope you agree it is unethical to use prisoners in that way.

(b) We have some sales literature that describes features of our product and compares it with several competing products. I think you might find this information helpful, so I'm sending you a copy.

(c) I know a number of their engineers because we went to college together. Believe me, they probably have copied our product as I doubt they would be skillful enough to design one from scratch.

(d) My boss says the overseas competitor will probably give you a low bid to get the initial work, but you will pay a lot more on any subsequent orders.

Solutions
Solution 9-1

As a consultant to the garbage company Jack needs to see that responsible management is aware of his liquid dumping discovery. The appropriate next step is to advise the company president. Neither (a) nor (b) are satisfactory as they fail to take appropriate action when some action is clearly called for. Answer (c), on the other hand, is a premature action that may be a public disclosure prior to a thorough evaluation of all the facts.
Answer is (d).

Solution 9-2
In purchasing equipment using a sealed bid method, Dave has the responsibility to treat all bidders fairly. If the sixth bidder indicated he could submit the bid by the morning following the stated bid deadline, this might be considered a reasonable request on this relatively small purchase. Thus Dave might consider it ethical and fair to allow this bidder a few additional hours to submit the quote. A substantial concession (like a weeks delay) would not seem appropriate unless it were made available to all bidders in a timely way. Opening and giving exact or approximate data to one bidder about competing sealed bids is unethical and unfair.
Answer is (c).

Solution 9-3

Alternative (c) is clearly inappropriate as it would not represent John's professional opinion. (a) and (d) might be suitable if John considered the chief engineer's desired allowable soil pressure value

obviously wrong. The (b) alternate seem most practical. Soil analysis is not an exact science. A reevaluation of the field test data might indicate a somewhat larger allowable soil pressure value-- possibly even the one the chief engineer wants. Thus the reevaluation and final report would be accurate and professional.
Answer is (b).

Solution 9-4

Jim immediately is faced with a situation where he must recognize what ethical conduct is. Alternatives (a) and (b) are unethical practices. Neither misrepresenting oneself to a competitor nor disclosing an employers confidential data are appropriate. Between alternatives (c) and (d) the situation is less clear. An ethical engineer must proceed forthrightly, but must he take the path of alternative (d)?

Jim's problem may be his inexperience. Other alternatives are available such as requesting all available sales literature of the competitor's products. By careful study, and by knowing the basic manufacturing and development capabilities, one can usually correctly predict the direction of products. Further, since this evaluation would naturally help decide the direction of your own product line, then a more careful market analysis of your own situation is certainly warranted. Often manufacturers purchase competitive products and benchmark its manufacturing technology. Since this is an ethics test question, you better choose to be on the safe side.
Answer is (d).

Solution 9-5

Answers (a), (c), and (d) represent malicious comments attempting to injure a competitors reputation. Answer (b) represents material that has been examined within Glenn's company and has been approved for distribution. Since it was intended for customers or prospective customers, Glenn would appear to be acting in an appropriate and ethical manner if he selects this alternative.
Answer is (b).

AFTERNOON SAMPLE EXAMINATION

Civil	No. Probs.
Soil Mechanics & foundations	6
Structural Analysis, frames, trusses	6
Hydraulics & hydro systems	6
Structural Design, concrete, steel	6
Environmental Engineering, waste water, solid waste treatment	6
Transportation facilities	6
Water purification and treatment	6
Surveying	6
Computer and Numerical methods	6
Legal & professional aspects	3
Construction Management	3

A-2

Instructions for Afternoon Session

1.

You have four hours to work on the afternoon session. You may use the *Fundamentals of Engineering Reference Handbook* as your *only* reference. Do not write in this handbook.

2.

Answer every question. There is no penalty for guessing.

3.

Work rapidly and use your time effectively. If you do not know the correct answer, skip it and return to it later.

4.

Some problems are presented in both metric and English units. Solve either problem.

5.

Mark your answer sheet carefully. Fill in the answer space completely. No marks on the workbook will be evaluated. Multiple answers receive no credit. If you make a mistake, erase completely.

Work 60 afternoon problems in four hours.

FUNDAMENTALS OF ENGINEERING EXAM

AFTERNOON SESSION

Ⓐ Ⓑ Ⓒ   Fill in the circle that matches your exam booklet

1 Ⓐ Ⓑ Ⓒ Ⓓ   16 Ⓐ Ⓑ Ⓒ Ⓓ   31 Ⓐ Ⓑ Ⓒ Ⓓ   46 Ⓐ Ⓑ Ⓒ Ⓓ

2 Ⓐ Ⓑ Ⓒ Ⓓ   17 Ⓐ Ⓑ Ⓒ Ⓓ   32 Ⓐ Ⓑ Ⓒ Ⓓ   47 Ⓐ Ⓑ Ⓒ Ⓓ

3 Ⓐ Ⓑ Ⓒ Ⓓ   18 Ⓐ Ⓑ Ⓒ Ⓓ   33 Ⓐ Ⓑ Ⓒ Ⓓ   48 Ⓐ Ⓑ Ⓒ Ⓓ

4 Ⓐ Ⓑ Ⓒ Ⓓ   19 Ⓐ Ⓑ Ⓒ Ⓓ   34 Ⓐ Ⓑ Ⓒ Ⓓ   49 Ⓐ Ⓑ Ⓒ Ⓓ

5 Ⓐ Ⓑ Ⓒ Ⓓ   20 Ⓐ Ⓑ Ⓒ Ⓓ   35 Ⓐ Ⓑ Ⓒ Ⓓ   50 Ⓐ Ⓑ Ⓒ Ⓓ

6 Ⓐ Ⓑ Ⓒ Ⓓ   21 Ⓐ Ⓑ Ⓒ Ⓓ   36 Ⓐ Ⓑ Ⓒ Ⓓ   51 Ⓐ Ⓑ Ⓒ Ⓓ

7 Ⓐ Ⓑ Ⓒ Ⓓ   22 Ⓐ Ⓑ Ⓒ Ⓓ   37 Ⓐ Ⓑ Ⓒ Ⓓ   52 Ⓐ Ⓑ Ⓒ Ⓓ

8 Ⓐ Ⓑ Ⓒ Ⓓ   23 Ⓐ Ⓑ Ⓒ Ⓓ   38 Ⓐ Ⓑ Ⓒ Ⓓ   53 Ⓐ Ⓑ Ⓒ Ⓓ

9 Ⓐ Ⓑ Ⓒ Ⓓ   24 Ⓐ Ⓑ Ⓒ Ⓓ   39 Ⓐ Ⓑ Ⓒ Ⓓ   54 Ⓐ Ⓑ Ⓒ Ⓓ

10 Ⓐ Ⓑ Ⓒ Ⓓ   25 Ⓐ Ⓑ Ⓒ Ⓓ   40 Ⓐ Ⓑ Ⓒ Ⓓ   55 Ⓐ Ⓑ Ⓒ Ⓓ

11 Ⓐ Ⓑ Ⓒ Ⓓ   26 Ⓐ Ⓑ Ⓒ Ⓓ   41 Ⓐ Ⓑ Ⓒ Ⓓ   56 Ⓐ Ⓑ Ⓒ Ⓓ

12 Ⓐ Ⓑ Ⓒ Ⓓ   27 Ⓐ Ⓑ Ⓒ Ⓓ   42 Ⓐ Ⓑ Ⓒ Ⓓ   57 Ⓐ Ⓑ Ⓒ Ⓓ

13 Ⓐ Ⓑ Ⓒ Ⓓ   28 Ⓐ Ⓑ Ⓒ Ⓓ   43 Ⓐ Ⓑ Ⓒ Ⓓ   58 Ⓐ Ⓑ Ⓒ Ⓓ

14 Ⓐ Ⓑ Ⓒ Ⓓ   29 Ⓐ Ⓑ Ⓒ Ⓓ   44 Ⓐ Ⓑ Ⓒ Ⓓ   59 Ⓐ Ⓑ Ⓒ Ⓓ

15 Ⓐ Ⓑ Ⓒ Ⓓ   30 Ⓐ Ⓑ Ⓒ Ⓓ   45 Ⓐ Ⓑ Ⓒ Ⓓ   60 Ⓐ Ⓑ Ⓒ Ⓓ

DO NOT WRITE IN BLANK AREAS

SAMPLE EXAM

1.
A soil specimen has a void ratio of 0.6, moisture content of 10%, and $G = 2.67$. Determine the moist unit weight.
(A) 13.4 kN/m³
(B) 15.6 kN/m³
(C) 18 kN/m³
(D) 20.1 kN/m³

2.
Calculate the seepage loss through the 0.5 m thick sandy silt layer below the earth dam shown (m³/hr/m). Given: coefficient of permeability, $k = 3.66$ cm/hr.
(A) 0.0023
(B) 0.0037
(C) 0.0051
(D) 0.0093

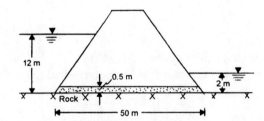

3.
A normally consolidated clay layer having a thickness of 3 m is shown. A laboratory consolidation test on the same clay gave the following results:

p (kN/m²)	e
75	0.0661
150	0.557

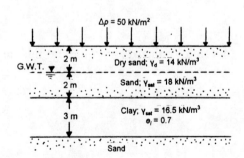

The settlement of the clay layer due to a surcharge $\Delta p = 50$ kN/m² is closest to:
(A) 0.14 m
(B) 0.25 m
(C) 0.56 m
(D) 0.75 m

4.

A silty clay layer of 6 m thickness is shown. It has a drained friction angle of $21°$ and a cohesion of 20 kN/m². What would be the drained shear strength at A (in kN/m²)?

(A) 30
(B) 45
(C) 60
(D) 75

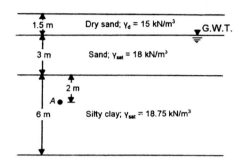

5.

For a granular soil, the dry unit weight in the field is 16.35 kN/m³. The maximum and minimum dry unit weights of the same soil as determined in the laboratory are 17.6 kN/m³ and 14.46 kN/m³, respectively. Given $G = 2.66$, the relative density of compaction in the field is closest to:

(A) 0.48
(B) 0.53
(C) 0.65
(D) 0.76

6.

For a clayey soil, given:
Plasticity index = 18
Plastic limit = 15
Specific gravity of soil solids, $G = 2.68$
Determine the void ratio of the soil at liquid limit. Note: At liquid limit, the soil is saturated.

(A) 0.60
(B) 0.70
(C) 0.80
(D) 0.90

7. The ratio of width b to depth y that maximizes the discharge through a fixed cross-sectional area in an open channel of rectangular section is

(A) b/y = 1/3
(B) b/y = 1/2
(C) b/y = 1
(D) b/y = 2

8. The depth of flow in a concrete-lined (n = 0.012) triangular channel, laid on a slope of 2m per 3km and carrying a discharge of 1.0 m³/s, is closest to

(A) 0.85 m
(B) 0.91 m
(C) 0.97 m
(D) 1.03 m

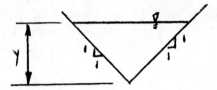

9. The hydraulic radius for a circular pipe of diameter D flowing half full is

(A) D/4
(B) D
(C) πD/2
(D) πD

10. This troweled concrete channel conveys water at a depth of one meter on a slope of 0.0004. The discharge is closest to

(A) 1.0 m³/s
(B) 2.1 m³/s
(C) 3.2 m³/s
(D) 10.0 m³/s

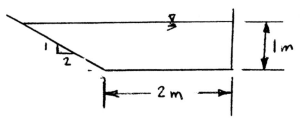

11. A 2-m-diameter pipe connecting two reservoirs has an interior surface of smooth masonry (C = 120) and is 5000 m long. The difference in reservoir water surface elevations is 6.0m. According to the Hazen-Williams formula, the discharge is nearest to

(A) 1.7 m³/s
(B) 2.7 m³/s
(C) 5.5 m³/s
(D) 8.5 m³/s

12.
A new cast iron pipe (C = 130) is to convey 4.5 m³/s a distance of 6 km from one reservoir to another reservoir whose water surface is 7.5m lower than the first reservoir. The most appropriate pipe diameter is

(A) 1.2 m
(B) 1.8m
(C) 2.4 m
(D) 3.0 m

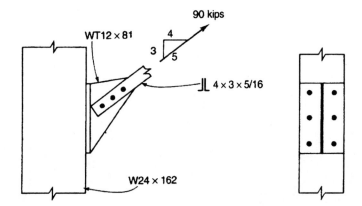

Problem 13 and 14.

The relevant properties of one angle are A_g = 2.09 inches², hole diameter = d_b + 0.125 inches. The allowable stress may be increased by one third when considering wind loads.

13. All bolts in the wind bracing connection shown in the figure are ⅞-inch A325 slip-critical bolts in standard holes.(A_g = 2.09) All structural sections are Grade A36 steel. The effective net area of the angle brace is most nearly:

(A) 2.0 square inches   (C) 3.0 square inches
(B) 2.5 square inches   (D) 3.5 square inches

14. The capacity of the angle brace based on its effective net area is most nearly:
     F_u = 58. A_e = 3.02

(A) 101 kips   (C) 111 kips
(B) 106 kips   (D) 116 kips

15. The W14x48 Grade A36 column shown below supports an axial load of 250 kips. Determine if the column is adequate. The relevant properties of the $W14 \times 48$ are A_s = 14.1 inches², E_s = 29,000 ksi, F_y = 36 ksi, r_x = 5.85 inches, r_y = 1.91 inches.

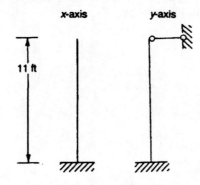

Figure for Prob. 15.

16. A 500 mm diameter, spirally reinforced column with 40 mm cover to the spiral is reinforced with ten No. M30 bars. The concrete strength is 35 MPa and the tensile strength of the reinforcement is 400 MPa. The column may be classified as a short column. $\phi P_n = 0.85 \, \phi \, [0.85 \, f'_c (A_g - A_{st}) + f_y A_{st}]$

The column supports only axial load and the design axial load strength is given most nearly by:

(A) 4800 kN   (C) 5200 kN
(B) 5000 kN   (D) 5400 kN

17. For the column in problem 16, the minimum permitted diameter of the spiral reinforcement is:

(A) ¼-inch   (C) ½-inch
(B) ⅜-inch   (D) ⅝-inch

18. For the column in problem 16, Is the reinforcement ratio of the column satisfactory?
    (A) yes   (B) no.

19. A 300 m radius simple curve with a tangent deflection of 40 degrees has been staked on the ground. It is desired to **sharpen** the curve so as to move the midpoint of the curve a distance of 3 m **radially**. The tangent deflection must remain fixed at 40 degrees so the direction of the tangents is not changed. The radius (m) of the new curve is most nearly:

    (A)   297
    (B)   331
    (C)   253
    (D)   340

20. An existing vertical curve joins a +4% with a -4% grade. If the length of the curve is 75 m, the maximum safe speed (km/h) on the curve is most nearly: (assume $f = 0.4$, a perception-reaction time of 2.5 sec., and that S<L)

    (A)   50
    (B)   40
    (C)   60
    (D)   45

Questions 21-22

Given the following traffic count data:

Time Interval	No. of Vehicles
5:00-5:15	1000
5:15-5:30	1100
5:30-5:45	1200
5:45-6:00	900

21. The peak hour factor is most nearly:

    (A)   0.950
    (B)   0.875
    (C)   1.050
    (D)   0.750

22. The peak rate of flow (veh/hr) is most nearly:

    (A)   3600
    (B)   4400
    (C)   4800
    (D)   3675

Questions 23-24

Given the following spiral and simple curve data: PI at station 1+086.271, R = 300 m, Δ = 16 degrees, ℓ_s = 52.083 m, p = 0.377 m, k = 26.035 m.

23. Assuming a rate of increase of centripetal force (C) of 0.6 m/s^3, the recommended design speed (km/h) for the curve is most nearly:

    (A)   100
    (B)   60
    (C)   120
    (D)   75

24. The stationing of the TS for the curve is most nearly:

    (A)   1+018
    (B)   1+154
    (C)   1+138
    (D)   1+000

25. A section of highway has the following design requirements.

Design speed = 80 km/hr
Coefficient of friction = 0.35
Perception-reaction time = 2.5 sec
Grade = +3%
Driver eye height = 1070 mm
Object height = 150 mm

The stopping sight distance (m) for this highway is most nearly:

    (A)   135
    (B)   120
    (C)   130
    (D)   110

26. The areas (in sq. ft.) of the cut sections of a proposed highway are shown below.

Stations	Area (sq. m.)
5+000	27.9
5+050	326.1
5+100	965.3
5+200	1651.8
5+300	2126.6

The total volume of cut (to the nearest cubic meter) between stations 5+000 and 5+100 is most nearly: (use the average end area method)

    (A)   35615
    (B)   34865
    (C)   44420
    (D)   41135

27. A 11-m wide vertical wall roadway tunnel is to be constructed using the cut and cover method. The cross sections to be excavated are shown below.

Station	Area (sq. m.)
1+000	20
1+100	30
1+130	35
1+200	40
1+260	45
1+300	40
1+400	35

The volume of earth in cubic meters to be excavated between stations 1+130 and 1+260 is most nearly: (use the prismoidal method)

    (A)   5165
    (B)   5585
    (C)   7860
    (D)   2340

28. The binary representation 10101 corresponds to which base 10 number?
    (A) 3
    (B) 16
    (C) 21
    (D) 10,101

29. For the base 10 number of 30, the equivalent binary number is
    (A) 1110
    (B) 0111
    (C) 1111
    (D) 11110

30. On personal computers the data storage device with the largest storage capacity is most likely to be
    (A) 3½" diskette
    (B) hard disk
    (C) random access memory
    (D) 3½" HD diskette

31. Pseudocode can best be described as
    (A) A simple letter-substitution method of encryption.
    (B) An English-like language representation of computer programming.
    (C) A relational operator in a database.
    (D) The way data are stored on a diskette.

32. All of the following is true of spreadsheets except:
    (A) A cell may contain label, value, formula or function.
    (B) A cell reference is made by a numbered column and lettered row reference.
    (C) A cell content that is displayed is the result of a formula entered in that cell.
    (D) Line graphs, bar graphs, stacked bar graphs and pie charts are typical graphs created from spreadsheets.

33. The following polynomial obviously has three roots:
    $$P(x) = x^3 + 4x^2 + 6x + 4$$
    Is $x = -1$ one of the exact roots?
    (A) No, because its less than the last coefficient.
    (B) Yes, because there is a remainder when using synthetic division.
    (C) No, because there is a remainder when using synthetic division.
    (D) Yes, there is no remainer.

34. For above problem if a second trial root to be found (using the Newton's method for finding roots) what is this next trial division?
    (A) 1
    (B) 2
    (C) -1
    (D) -2

35. For a differential equation: $dx/dt + 5x = 1$, $x(0)=0$; What is the value of x suggested by the Euler's method after first iteration after initial condition?
       (A) 0.1
       (B) 0.2
       (C) 0.3
       (D) 0.4

36. The *NCEES Model Rules of Professional Conduct* do not allow an engineer to improve his/her competence in a specific area in which one of the following ways?
(A)   Studying the work of other engineers.
(B)   Study at a university outside of the United States.
(C)   Performing services in an area where the engineer is not yet competent.
(D)   Work in a specific area under the direction of a competent but unregistered engineer.

37. The *NCEES Model Rules of Professional Conduct* requires the registered engineer to do all but one of the following actions. Which one should not be done?
(A)   Make no statements on technical matters unless any interested party for whom the engineer is acting is disclosed.
(B)   Approve documents only if they safeguard the life, health, property, and welfare of the public.
(C)   Reveal facts and data obtained in a professional capacity only after obtaining consent of the client or employer, unless required to reveal the facts and data by law.
(D)   Comment on professional matters publicly even though the engineer has not yet had an opportunity to make a careful evaluation of the facts and subject matter.

38. The *NCEES Model Rules of Professional Conduct* prohibits the registered engineer from doing all but one of the following. Which one can the engineer do?
(A)   Indiscriminately criticizing the work of another registered engineer.
(B)   Giving gifts to people in order to promote a business and obtaining work.
(C)   Affixing a signature or seal to any plans or documents not prepared under the engineers control and supervision.
(D)   Accepting compensation from more than one party for services on the same project, even
        though the circumstances are fully disclosed and agreed to by all parties.

39. Jane Doe is a self-employed consulting engineer in the engineering department of ABC company where you work as a full time staff process engineer. One of your expertises is relief system design. Things are slow, and even though you are well recognized, you are facing a potential layoff. She approached you for part time help, which may be summarized by the following conversation:
Jane: "Hey Phil, how about earning some extra money, say forty dollars per hour, from XYZ Company, for helping in my calculations to check the size of relief devices on weekends? I will do all the leg work, collect all the pertinent data, all you have to do is to check whether the device is OK or not OK, and show the basis of your calculations."
Phil: "Is it not illegal to have a part time job when I am a full time employee?"

Jane: "I don't think so. I know John, your project manager, has a restaurant business. Roger, your own colleague, helps in income tax preparation on weekends."
What would you do?

(A) Check the employment agreement, and make sure nothing was signed against accepting a part-time job, and accept the offer and work at home without using any company time or resources, and follow "don't-ask-don't-tell" principle.
(B) Accept the offer, and work at the office outside working hours, and use the company's methods and copier.
(C) Read the company ethics book and guide lines to check whether there is any clear definition of Conflict of Interest. If the definition is clear, reject the offer. If unclear, ask the Human Resources Department for permission by a short memo, and then decide.
(D) Invite Jane and your boss for a dinner, and casually refer to the situation at an appropriate moment, and have his permission well heard so that Jane remains a witness, and then accept the offer.

40. Suppose you work for a large electrical design firm that designs electrical control systems for industrial manufacturing sites. Assume your design requirements include specifying "short circuit and ground fault protection" for motor circuits. These device specifications should comply with the National Electric Code. Your design, started in early 1991, involved the selection of a non-time delay fuse for a three-phase, 230V, 50 hp, 115A (it's name plate values) induction motor for a molded rubber parts plant. The code book states that the full rated current for this motor is to be 130 amperes. The book also states that for this type of non-time delay fuse the current should not be more than 300% of rated. However, you find that commercially available fuses come only in increments of 50 amperes near that value. You choose the 400 amp fuses since the code implies that is permissible (and is a widely accepted practice) to "round up" to this higher value. Your design was not implemented as the project was tentatively canceled due to lack of funding. In 1995 the exact same project was funded and the project went ahead. Assume you have gone on to work on other projects. You are aware that in the 1993 edition of this handbook a number of revisions were made, one requiring fusing and circuit breakers to use the next lower, or "rounding down", of the fuse and circuit breaker sizes. What action, if any should you take?

(A) Ignore the project since you are working in another area now.
(B) Call the electrical engineer you know is working on the project reminding him that the prior work was based on the code in force in 1991 and not the latest code.
(C) Write a memo to the project supervisor of the change in the code and the need to revise the fuse rating.
(D) The client for the project is that firm that produces molded rubber parts. They have a project manager overseeing the plant design. Call this project manager and tell him of the needed change in fuse rating to match the present electrical code.

41. A certain "hi-tech" company, located in California, produces a very popular electronic device that emits an amount of radiation that is within established limits of radiation. The company sells all of the product that it produces, both domestically and abroad. The company is able to increase its production by buying some of the required components overseas. You are the supervisor of the company's test laboratory which samples and tests for radiation levels. It is a standard and acceptable practice to sample a certain percentage of the devices produced to check for compliance. After a large production run, your sampling procedure shows that the limits of radiation have been exceeded. You advise your company's managers of this higher radiation level due to the variation of overseas components (that is, their quality control is poor). A meeting is called consisting of yourself, two other engineers, and two plant managers; the cause of the problem, of course, is the variation in the overseas components. The possible courses of action discussed are:

(A) The tolerances of a few of the sampled components are usually detectable. To create a distorted testing record, selectively choose the samples to test. Because the test records will show the percent sampled components to be within limits, ignore the problem (there are not that many devices too far out of specs).
(B) Curtail the sales in the United States and sell to the overseas countries that don't have the radiation requirements.
(C) Stop all sales until a new supplier is found for higher quality components; and, attempt to modify existing devices to meet sample specifications. This action may delay an impeding cost of living salary increase for most of the plant's employees.
(D)Remove the label on each of the devices that states that the device meets all radiation requirements, reduce the price to remove the inventory, then phase in newer devices made with higher quality components.

Each member of the committee is highly respected for his judgement, the managements wants honest opinions. Four of the members have voted, one for each of the possible courses of action, your vote then will be the deciding one. Which will you vote for?

42-43.
A water has the following ionic composition:

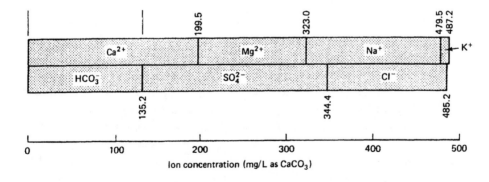

Ion concentration (mg/L as $CaCO_3$)

42. The total carbonate hardness as $CaCO_3$ is:
   (A) less than 100 mg/L.
   (B) Greater than the non-carbonate hardness.
   (C) only associated with Ca++.
   (D) 244.4 mg/L.

43. The $CA(OH)_2$ needed to remove the carbonate hardness (CH) is:

(A) 1 mole/mole CH.

(B) 2.2 mg/mg CH.

(C) 0.46 mg/mg CH.

(D) dependent upon both the calcium and magnesium concentrations.

44. A single source of BOD to a river causes a minimum DO downstream of 6 mg/L. The initial DO deficit is zero and DO saturation is 10 mg/L.

(A) without aeration, the deficit will decrease

(B) doubling the BOD load would double the deficit.

(C) Doubling the BOD load would reduce the DO to 3 mg/L.

(D) doubling the BOD would move the point where the deficit occurs closer to the discharge.

45. A distilled water solution has 0.001 M H_2SO_4. The pH is:

(A) 2.7

(B) 11.3

(C) neutral

(D) 3.0

46. Magnesium in equilibrium with precipitated magnesium carbonate is:

1. dependent upon pH.

2. dependent upon carbonate concentration

3. Dependent upon bicarbonate concentration

4. Dependent upon alkalinity

Which of the above statements are correct?

(A) Statements 1 and 4 are correct.

(B) Statements 2 and 3 are correct.

(C) Statements 1 and 2 are correct.

(D) all of the statements are correct.

47. In using Chlorine for disinfection:

(A) concentration of HOCl is pH dependent

(B) [HOCl] equals [OCl⁻] at pH 7

(C) effectiveness increases with time of contact.

(D) total elimination of bacteria is possible.

48. A 61 cm diameter circular concrete sewer is set on a slope of 0.001. The flow when it is half full is:

(A) ½ of flow at Q_{full}

(B) 0.08 m³/s

(C) 0.20 m³/s

(D) 0.28 m³/s

49. The carbonaceous oxygen demand of 200 mg/l of glycine $(CH_2(NH_2)(OOH)$ is:
   (A) less than the nitrogenous oxygen demand
   (B) depends on the reaction rate
   (C) greater than 120 mg/l
   (D) greater than 100 mg/l

50. A completely-mixed lagoon has a waste flow of 1 m³/s of a BOD waste with a decay coefficient of 0.3/d and a concentration of 100 mg/l. If the effluent BOD must be 20 mg/l or less, then:

   (A) Volume required is about 1×10^6 m³
   (B) Volume required is about 2×10^6 m³
   (C) Volume required is about 0.5×10^6 m³
   (D) An increase in temperature would increase the volume required.

51. A completely-mixed reactor with cell recycle is designed to treat a municipal waste. Assuming removal kinetics as in the equation below, then:

$$r_{su} = -\frac{kXS}{(K_s + S)}$$

   (A) the effluent substrate concentration decreases with greater θ_c.
   (B) The food:microorganism level in independent of θ_c.
   (C) Microbe concentrations will be smaller than the no-recycle case.
   (D) θ_c is independent of effluent quality.

52. Chlorination of wastewater effluents:
   1. Requires more chlorine than chlorination of drinking water.
   2. Requires 3 moles of chlorine for each mole of ammonia
   3. Is used to improve effluent quality
   4. Oxidizes other reduced chemicals such as ferrous iron.
Which of the above statments are correct?
   (A) All of the above statments are correct.
   (B) None of the above statments are correct.
   (C) 1 and 3 only are correct.
   (D) 1, 3, and 4 are correct.

53. A chemical is removed in a completely-mixed reactor at a zero order rate of 0.25 mg/l-d. The influent concentration is 10 mg/l and the hydraulic detention time is 10 days. For this reactor:
1. The effluent concentration would be the same as a batch reactor with a 2 day retention time.

2. The effluenmt concentration will be about 7.5 mg/l
3. The effluent quality would improve with longer detention times
4. The effluent quality would improve with increased mixing.
Which of the above statments are:
    (A) All of the above statments are correct.
    (B) None of the above statments are correct.
    (C) 1 and 2 only are correct.
    (D) 1, 3, and 4 are correct.

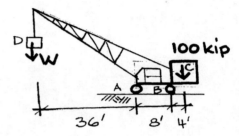

54. The load, W, at which the crane will tip is
    (A) more than 33.3 kips
    (B) more than 40 kips
    (C) more than 9.5 but less than 33.3 kips
    (D) none of the above

Prob. 54.

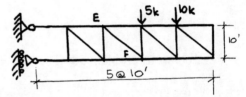

55. The force in bar EF is
    (A) 21.2 kips compressive
    (B) 21.2 kips tensile
    (C) 10.6 kips tensile
    (D) 10.6 kips compressive

Prob. 55.

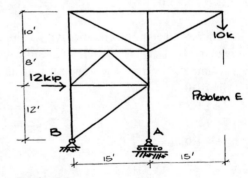

56. The reaction at B is
    (A) 23 kips up and to left
    (B) 23 kips up and to right
    (C) 23 kips down and to left
    (D) 23 kips down and to right

Prob. 56.

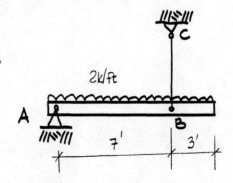

57. If the force in bar BC is 28.57 kips tensile,
    the reaction at A must be:
        (A) 8.6 kips down
        (B) 8.6 kips up
        (C) 18.6 kips up
        (D) 18.6 kips down

Prob 57.

58. The shearing force at section A-A is (ignore thickness of members)
    (A) 12.8 k
    (B) 2.8 k
    (C) 5.0 k
    (D) none of these

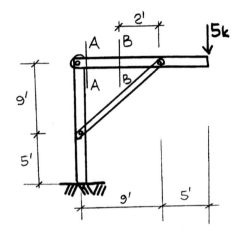

Prob. 58-59.

59. Same structure as problem 58. The bending moment at section B-B is:
    (A) 50.5 kft
    (B) 19.4 kft
    (C) 12.0 kft
    (D) 15.4 kft

60. A concrete lined open channel with a slope of 0.34% has a bottom width of 1.5 m and side sloped of 2:1. This channel is to be designed to handle a peak discharge of 15 m^3/s. Assume n = 0.015. The depth of flow in feet is nearest to:
    (A) 1.0 m
    (B) 1.2 m
    (C) 1.4 m
    (D) 1.6 m

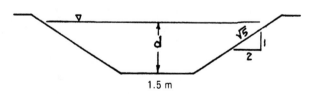

Prob. 60.

End of Exam. Please check your work.

SOLUTIONS

1. **Dry unit weight,**

    $$\gamma_d = \frac{G\gamma_w}{1+e} = \frac{(2.67)(9.81)}{1+0.6} = 16.37 \text{ kN}/\text{m}^3$$

    Moist unit weight,

    $$\gamma = \gamma_d(1+w) = (16.37)(1+0.1) = 18 \text{ kN}/\text{m}^3$$

    Answer is (C).

2. $Q = kiA$

    $k = 3.66 \text{ cm}/\text{hr}; \ i = \frac{12-2}{50} = 0.2; \ A = (0.5)(1) = 0.5 \text{ m}^2$

    $$Q = \left(\frac{3.66}{100}\right)(0.2)(0.5) = 0.0037 \text{ m}^3/\text{hr}/\text{m}$$

    Answer is (B).

3.
    $$C_c = \frac{e_1 - e_2}{\log\left(\dfrac{p_2}{p_1}\right)} = \frac{0.661 - 0.577}{\log\left(\dfrac{150}{70}\right)} = 0.279$$

    Effective overburden pressure at the middle of the clay layer,

    $$p_i = (14 \times 2) + (18 - 9.81)(2) + \frac{3}{2}(16.5 - 9.18) = 54.42 \text{ kN}/\text{m}^2$$

    $$\Delta H = \frac{C_c H}{1+e_i} \log\left(\frac{p_i + \Delta p}{p_i}\right) = \frac{(0.275)(3)}{1+0.7} \log\left(\frac{54.42 + 50}{54.42}\right) = 0.139 \text{ m} = 139 \text{ mm}$$

    Answer is (A).

4. $s = c + \sigma' \tan \phi$

    $c = 20 \text{ kN/m}^2; \ \phi = 21; \ \sigma' = (15)(1.5) + 3(18 - 9.81) + 2(18.75 - 9.81) = 64.95 \text{ kN/m}^2$

    $s = 20 + 64.95 \tan 21° = 44.9 \text{ kN/m}^2$

    Answer is (B).

5.

$$D_d = \dfrac{\dfrac{1}{\gamma_{min}} - \dfrac{1}{\gamma_d}}{\dfrac{1}{\gamma_{min}} - \dfrac{1}{\gamma_{max}}} = \dfrac{\dfrac{1}{14.46} - \dfrac{1}{16.35}}{\dfrac{1}{14.46} - \dfrac{1}{17.6}} = 0.645 = 64.5\%$$

Answer is (C).

6. $LL = PL + PI = 15 + 18 = 33.$ **Degree of saturation**

$$S(\%) = \frac{wG}{e} \times 100$$

If $S = 100\%$

$$e_{LL} = wG = \left(\frac{30}{100}\right)(2.68) = 0.804$$

Answer is (C).

7.

The Manning discharge formula is

$$Q = \frac{1}{n}\, AR^{\,2/3}S^{1/2}$$

For a fixed area A, the discharge is maximized when R is maximized. With A = by and the wetted perimeter P = b + 2 y,

$$R = \frac{A}{P} = \frac{by}{b+2y} = \frac{A}{\dfrac{A}{y}+2y}$$

$$\frac{dR}{dy} = \frac{-A}{\left(\dfrac{A}{y}+2y\right)^2}\left(-\frac{A}{y^2}+2\right) = 0$$

or A = $2y^2$ = by   so   b = 2y,   b/y = 2. The answer is (D).

8.

The discharge is

$$Q = \frac{1}{n}\, AR^{2/3}S^{1/2}$$

in which $A = y^2$

$$P = 2\sqrt{2}\,y$$

and $\qquad R = A/P = y/\left(2\sqrt{2}\right)$

Hence

$$1.0 = \frac{1}{0.012}\,y^2\left(\frac{y}{2\sqrt{2}}\right)^{2/3}\left(\frac{2}{3000}\right)^{1/2}$$

or $\quad y^{8/3} = 0.930$, which yields $\quad y = (0.930)^{3/8} = 0.973$ m. The answer is (C).

9.
The hydraulic radius is $R = A/P$. Using $r = D/2$,
for a half-full circular pipe one has $A = \pi r^2/2,\ P = \pi r$ so that

$$R = \frac{A}{P} = \frac{\pi r^2 / 2}{\pi r} = \frac{r}{2} = \frac{D}{4}$$

The answer is (a).

10.
The discharge is

$$Q = \frac{1}{n} AR^{2/3}S^{1/2}$$

From the table of Manning n values, n = 0.013. Also,

$$A = 2 + \frac{1}{2}(1)(2) = 3.0 \ \text{m}^2$$

$$P = 1 + 2 + \sqrt{5} = 5.24 \ \text{m}$$

Therefore $\quad Q = \frac{1}{0.013}(3.0)\left(\frac{3.0}{5.24}\right)^{2/3}(0.0004)^{1/2}$

and $\qquad Q = 3.2 \ \text{m}^3/\text{sec}$

The answer is (C).

11.

For smooth masonry the Hazen-Williams coefficient is C = 120. The Hazen-Williams formula is

$$V = 0.849CR^{0.63}S^{0.54}$$

Here

$$A = \frac{\pi}{4}D^2 = \frac{\pi}{4}(2)^2 = \pi \ m^2$$

$$P = \pi D = 2\pi \ m$$

$$R = A/P = 0.5 \ m$$

Neglecting local losses, $S = h_L/L = 6/5000$
Hence

$$V = 0.849(120)(0.5)^{0.63}\left(\frac{6}{5000}\right)^{0.54} = 1.74 \ m/s$$

and $\qquad Q = AV = \pi(1.74) = 5.47 \ m^3/s \qquad$ The answer is (C).

12.

The Hazen-Williams equation may be written as

$$V = 0.849CAR^{0.63}S^{0.54}$$

or as

$$AR^{0.63} = \frac{Q}{0.849CS^{0.54}} = \frac{4.5}{0.849(130)\left[\dfrac{7.5}{6000}\right]^{0.54}} = 1.507$$

In this case

$$A = \frac{\pi}{4}D^2 \quad \text{and} \quad R = A/P = D/4 \quad \text{for a pipe flowing full.}$$

Thus $\qquad \dfrac{\pi}{4}D^2\dfrac{D^{0.63}}{4^{0.63}} = 1.507$

$$D^{2.63} = \frac{1.507(4^{1.63})}{\pi} = 4.59$$

and

$$D = (4.59)^{1/2.63} = (4.59)^{0.38} = 1.785 \ m$$

The answer is (B).

13. From AISC Section B2, the net area of the brace is $A_n = A_g - 2\, td_h = 2$ x $2.09 - 2$ x 0.3125 $(0.875 + 0.125) = 3.56$ square inches. From AISC Equation (B3-1), the effective net area of the brace is $A_e = UA_n = 0.85$ x $3.56 = 3.02$ square inches. The answer is (C).

14. In accordance with AISC Section A5.2 a one-third increase in allowable stress is permitted for wind loading. The strength of the brace based on the effective net area, in accordance with AISC Section D1, is $P_t = 1.33$ x $0.5\, F_u A_e = 1.33$ x 0.5 x 58 x $3.02 = 116.48$ kips. The answer is (D).

15. The relevant properties of the $W14$ x 48 are $A_s = 14.1$ inches^2, $r_x = 5.85$ inches, $r_y = 1.91$ inches. The axial stress from the applied load is $f_a = P/A = 250/14.1 = 17.73$ kips per square inch. The slenderness ratio about the x-axis is given by $Kl/r_x = 2.1$ x 11 x $12/5.85 = 47.4$. The slenderness ratio about the y-axis is given by $Kl/r_y = 0.8$ x 11 x $12/1.91 = 55.3$, which governs. $C_c = (2\pi^2$ x $29{,}000/36)^{0.5} = 126.1 > Kl/r_y$. The allowable stress is
$F_a = [1 - 55.3^2 / (2$ x $126.1^2)]36/[5/3 + 3$ x $55/3 /(8$ x $126.1) - 55.3^3 / (8$ x $126.1^3)]$
   $= 17.88$ kips per square inch.
   > 17.73 Hence, the column is adequate.

16. For a spirally reinforced column, the design axial load strength is given by the equation as
   $\phi P_n = 0.85\ \phi\ [0.85\, f'_c(A_g - A_{st}) + f_y A_{st}] = 0.85$ x $0.75[0.85$ x$35(196{,}350 - 7000) + 400$ x $7000]$
   $= 5376$ kN. The correct answer is (D).

17. In accordance with ACI Section 7.10.4.2, the minimum diameter of the spiral reinforcement is ⅜-inch. The answer is (B).

18. In accordance with ACI Section 10.91, the minimum reinforcement ratio required is 0.01 and the maximum allowable reinforcement ratio is 0.08. The correct answer is (A).

19. The relationship between the external ordinate (E) and the radius (R) is:
   $E = [R/\cos(\Delta/2)] - R$
The external ordinate of the existing curve is:
   $E = [300/\cos(40/2)] - 300 = 19.25$ m
The new curve is to be shifted radially a distance of 3 m. Therefore, the external ordinate of the new curve is $19.25 - 3 = 16.25$ m. The new radius is:
$R = E/[(1/\cos \Delta/2) - 1] = 16.25/[(1/\cos 40/2) - 1] = 253.20$ m
The answer is (C).

20. Determine the stopping sight distance for the 75 m crest vertical curve.
   $L = (AS^2)/404$
   $75 = (8S^2)/404$
   $S = 61.54$ m
Determine the maximum safe speed for this sight distance.
   $S = 0.278Vt + V^2/[254 (f \pm g)]$

$61.54 = 0.278(V)(2.5) + V^2/[254\,(0.40 - 0.04)]$
From the quadratic equation:
$V = 49.69$ km/h
The answer is (A).

21.
$PHF = V_h/(4V_{15})$

where V_h = hourly volume (vph), V_{15} = maximum 15-minute rate of flow within the hour (veh).
$PHF = (1000 + 1100 + 1200 + 900)/(4 \times 1200) = 0.875$
The answer is (B).

22.
Peak flow rate = 4 x 1200 = 4800 veh/hr
The answer is (C).

23.
The relationship between speed and length of spiral is:
$\ell_s = V^3/(46.7RC)$
$52.083 = V^3/[(46.7)(300)(0.6)]$
$V = 75.9$ km/h
The answer is (D).

24.
Compute the spiral tangent.
$T_s = (R + p)\tan(\Delta/2) + k = (300 + 0.377)\tan(16/2) + 26.035 = 68.250$ m
TS station = PI station - T_s = 1+086.271 - 0+068.250 = 1+018.021
The answer is (A).

25.
Stopping sight distance (S) is:
$S = 0.278Vt + V^2/[254(f \pm g)] = 0.278(80)(2.5) + (80)^2/[254(0.35 + .03)] = 121.9$ m
The answer is (B).

26.

Stations	Area (m^2)	Total area	Average Area	Distance (m)	Volume (m^3)[a]
5+000	27.9				
		354	177.0	50	8850.0
5+050	326.1				
		1291.4	645.7	50	32285.0
5+100	965.3				
				Total Volume =	41135.0

[a] Volume = distance x average area

The answer is (D).

27.

The bases of the cross sections are constant (11m.). The heights of the cross sections vary between stations but can be determined by dividing the cross section areas by 11m. The prismoidal formula is:

$$V = (L/6)(A_1 + 4A_m + A_2)$$

Determine the volume between stations 1+130 and 1+200.

$A_m = [(3.2 + 3.6)/2]11 = 37.4$ sq. m.

$V = (70/6)[35 + 4(37.4) + 40] = 2620.3$ m^3

Determine the volume between stations 1+200 and 1+260.

$A_m = [(3.6 + 4.1)/2]11 = 42.4$ sq. m.

$V = (60/6)[40 + 4(42.4) + 45] = 2546.0$ m^3

Total volume = 2620.3 + 2546.0 = 5166.3 m^3

The answer is (A).

28.

The binary (base 2) number system representation is

	2^4	2^3	2^2	2^1	2^0
	16	8	4	2	1
Binary number	1	0	1	0	1
	16	0	4	0	1 = 21

Answer is (C).

29.

2^5	2^4	2^3	2^2	2^1	2^0
32	16	8	4	2	1
1	1	1	1	0	= 30

Answer is (D).

30. Answer is (B).

31. Answer is (B).

32.

A cell reference is made by a lettered column and a numbered row, example C3.

Answer is (B).

33. Synthetic division

```
    1    4    6    4  |-1   trial root
        -1   -3   -3
    1    3    3  |1  a remainder of 1
```

Answer is (B).

34.

$x^{n+1} = x_n - P(x_n)/P'(x_n)$
for $x_1 = -1$, $P'(x) = 3x^2 + 8x + 6 \mid_{x=-1} = 3 - 8 + 6 = 1$
$P(x)\mid_{x=-1} = -1 + 4 - 6 + 4 = +1$
$\therefore x_{n+1} = (-1) - (1/1) = -2$

Answer is (D).

35.
Using equation 4.9a

$(x_{k+1}) = x_k - Tax_k + Tf_k$ for fixed increments.
$x_1 = x_0 - Ta_{x0} + Tf = 0 - 0 + 0.3 \times 1 = 0.3$

Answer is (C).

36. The answers (A), (B), and (D) represent practical and appropriate ways of improving one's competence. Answer (C) however may subject the client or employer to work by an engineer who is not fully competent in the specific area. This is clearly a less appropriate way to gain competency in a technical area.
Answer is (C).

37. Answers (A), (B), and (C) are prescribed in the *Model Rules*. Section Ie states that engineer's shall express a professional opinion publicly when it is based on adequate knowledge and a competent evaluation. Answer (D) is commenting publicly *before* a careful evaluation.
Answer is (D).

38. Professional conduct precludes an engineer from doing the things in (A), (B), and (C). Section IIg says the engineer shall not accept compensation from more than one party for services on the same project unless the circumstances are fully disclosed and agreed to by all parties. Thus the conduct in (D) is acceptable.
Answer is (D).

39. The best answer is (C). If (A) is chosen and discovered, you may jeopardize your employment. Working at the office, using company resources, tools, or proprietary methods, would be in conflict of interest, so answer (B) must be excluded. Answer (D) might be acceptable but relies on involved parties as witnesses to an agreement.

40. As a professional engineer you have a responsibility to safeguard life, health, and property. Thus you should take appropriate action to advise the proper person or people of the need to review the fuse selection. The question becomes who should one contact. While alternative (B) might be all that is needed, Alternative (C) is a more formal action (memo instead of a call), is addressed to the supervisor level, and is therefore preferred. Alternative (D) is premature since you do not know the current state of the design (the fuse size may already have been changed). Almost certainly (D) is not acceptable to your employer or in accordance with his policies on how to handle matters like this.

Answer is (C).

41. Although, in a legal sense, only action (A) is unacceptable; items (B) and (D) may possibly not be in the best interest of everyone because of unknown human effects, therefore item (C) is in the best interest of public safety. Answer is (C).

42. Answer is (C).

43. Answer is (A).

44. Answer is (B).

45. Answer is (A).

46. Answer is (D). All of the statments are correct.

47. Answer is (A).

48. $Q_f = 0.2$ m^3/s
$Q/Q_f = 0.4$
$Q = 0.08$ m^3/s
Answer is (B).

49. Answer is (C).

50.

$$\frac{A}{A_0} = \frac{1}{1 + K_1 \frac{v}{Q}}$$

$$0.2 = \frac{1}{1 + K_1 \frac{v}{Q}}$$

$$1 + K_1 \frac{v}{Q} = 5$$

$$v = \frac{4Q}{K_1} = \frac{3.46 \times 10^5}{.3} = 1.1 \times 10^6 \, m^3$$

Answer is (A).

51. Answer is (A).

52. Answer is (D).

53.
(C).

Answer is

54.

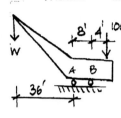

(a) Tips about A when $R_B = 0$

ΣM_A ⟳

$-(W)(36) + 100k(12) = 0$

$W = \frac{100}{3} = \underline{33.3k}$

If $W > 33.3$ will tip about A

(b) Tips about B when $R_A = 0$

ΣM_B ⟳

$-W(44) + 100(4) = 0$

$W = 400/44 = \frac{400}{42} = 9.09k$ 9.5k

Hence W must lie between 9.09 and 33.3k. Answer is (c).

55.

Use FBD shown

$\Sigma V = 0$ ↑+

$-5 - 10 + 0.707(EF) = 0$

$EF = +21.2$

assumed dir^n ok so $EF = \underline{21.2k}$ t

56.

$\Sigma H = 0$ gives R_B (horiz comp) = ← 12k

$\Sigma M_A = 0$ gives R_B (vert comp) = ↓ 19.6

Resultant ≈ ↙ 23k

57.

From $\Sigma H = 0$ $H_A = 0$

$\Sigma V = 0$ ↓+ gives (FBD of beam AB)

$+28.57 - 2 \times 10 + V_A(↑) = 0$

$V_A(↑) = 8.57$ best ans is 8.6

58.

(a) FBD whole struct.

$\Sigma H :$ H_A = zero

$\Sigma V :$ $V_A = 5 \uparrow$

$\Sigma M :$ $M_A = 70\,kft$ ccw.

(b) FBD

ΣM_A gives $v = 70/9 = 7.77$

Since strut is @ 45° $v = h$

Hence Shear @ AA $= 7.77 - 5.00 \cong \underline{2.8k}$.

59.

ΣM_{BB} ↻

$+5(7) - 7.77(2) + M_{BB} = 0$

$M_{BB} = -(35 - 15.54) = \underline{-19.4\,kft}$

60. Find the flow depth, d, in feet.

Manning's Formula

$$V = \frac{1}{n} R^{2/3} S^{1/2}$$

where V = velocity

$R = A / P$

A = channel cross-sectional area

P = Wetted perimeter

$$Q = AV = \frac{1}{n} A R^{2/3} S^{1/2}$$

For this trapezoidal channel

$A = (1.5 + 2d)d$ $P = 1.5 + 2\sqrt{5}d$

Hence

$$15 = \frac{1}{0.015}(1.5 + 2d)d\left[\frac{(1.5 + 2d)d}{1.5 + 2\sqrt{5}d}\right]^{2/3}(0.0034)^{1/2}$$

Solve by trial for d; i.e., choose a value of d, substitute into the equation and see whether it is satisfied.

We find d = 1.23m, which gives A = 4.87 m^2.

Answer is (B).

Worksheet for problem 60.

$AR^{2/3} = 3.8587$ Where $A=(1.5 +2d)d$
                                           $P=1.5 + 2\sqrt{5}\, d$
                                           $R=A/P$

d	A	P	R	$R^{2/3}$	$AR^{2/3} \stackrel{?}{=} 3.8587$
1.0	3.50	5.97	0.586	0.701	2.45
1.5	6.75	8.21	0.822	0.878	5.92
1.2	4.68	6.87	0.681	0.774	3.62
1.25	5.00	7.09	0.705	0.792	3.96
1.23	4.87	7.00	0.696	0.785	3.82

$y = 1.23$ m

Engineer-in-Training License Review

Donald G. Newnan, Ph.D., P.E., Civil Engr., Editor

A Complete Textbook Prepared Specifically For The New Closed-Book Exam

Written By Ten Professors - Each An Expert In His Field
* 220 Example Problems With Detailed Solutions
* 518 Multiple-Choice Problems With Detailed Solutions
* 180-Problem Sample Exam - The 8-hour Exam
    With Detailed Step-by-Step Solutions
* Printed In Two Colors For Quicker Understanding Of The Content

Here is a new book specifically written for the closed book EIT/FE exam. This is not an old book reworked to look new - this is the real thing! Ten professors combined their efforts so each chapter is written by an expert in the field. These include engineering textbook authors: Lawrence H. Van Vlack (*Elements of Materials Science and Engineering*), Charles E. Smith (*Statics, Dynamics, More Dynamics*) and Donald G. Newnan (*Engineering Economic Analysis*).

The Introduction describes the exam, its structure, exam-day strategies, exam scoring, and passing rate statistics. The exam topics covered in subsequent chapters are Mathematics, Statics, Dynamics, Mechanics of Materials, Fluid Mechanics, Thermodynamics, Electrical Circuits, Materials Engineering, Chemistry, Computers, Ethics and Engineering Economy. The topics carefully match the structure of the closed-book exam and the notation of the *Fundamentals of Engineering Reference Handbook.* Each topic is presented along with example problems to illustrate the technical presentation. Multiple-choice practice problems with detailed solutions are at the end of each chapter.

There is a complete eight-hour EIT/FE Sample Exam, with 120 multiple-choice morning problems and 60 multiple-choice general afternoon problems. Following the sample exam, complete step-by-step solutions are given for each of the 180 problems.

60% Text. 40% Problems & Solutions. 8-1/2 x 11, 2-Color,
ISBN 1-57645-001-5   5.0 lbs   Yellow/Green **Hardbound** Cover
Order No. 015   784 Pgs 14th Ed 1996 $39.95

EIT Civil Review

Donald G. Newnan, P.E., Civil Engineer, Ph.D., Editor
Written by Six Civil Engineering Professors for the closed book afternoon FE/Civil Examination.
Each topic is briefly reviewed with example problems. Many end-of-chapter
problems with complete step-by-step solutions and a complete afternoon
Sample Exam with solutions are included.
Soil Mechanics and Foundations
Structural Analysis, frames, trusses
Hydraulics and hydro systems
Structural Design, Concrete, Steel
Environmental Engineering, Waste water
Solid waste treatment
Transportation facilities
Water purification and treatment
Computer and Numerical methods ISBN 1-57645-013-9

Order No. 139   1998   8-1/2 x 11 $29.95

Civil Engineering License Review

Donald G. Newnan, Ph.D., P.E., Civil Engineer, Editor

Written by Six Professors, each with a Ph.D. in Civil Engineering.

A **complete license review book** written for the National Civil Engineering/Professional engineer exam.

☆ The 12 sections cover all the essential materials on the exam.
☆ A detailed discussion of the exam and how to prepare for it!
☆ 335 exam essay and multiple-choice problems with a total of 650 individual questions
☆ A complete 24 problem Sample Exam
☆ Printed in two colors for quicker understanding of the content

This is a new license review book written specifically for the latest version of the Civil Engineering/Professional Engineer (principles and practice) exam. The 12 sections cover the topics on the exam: Buildings, Bridges, Foundations, Retaining Structures, Seismic Design, Hydraulics, Engineering Hydrology, Water Treatment, Wastewater Treatment, Geotechnical/Soils, and Transportation Engineering. In addition, an appendix discusses Engineering Economy, which may appear as a component in some problems. A complete 24-problem sample exam (12 essay and 12 multiple-choice problems) is the final chapter in the book. Since some states do not allow books containing solutions to be taken into the CE/PE exam, the end-of-chapter problems do not have the solutions in this book. Instead, the problems are restated together with detailed solutions in the companion *book: Civil Engineering License Problems and Solutions* listed below. 80% Text 20% Problems. 2 Color **ISBN 0-910554-90-0** Hardbound Yellow/Blue 8-1/2 x 11
Order No. 900 728 Pgs 1995 12th Edition $51.50

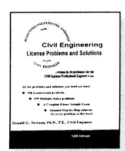

Civil Engineering License Problems and Solutions

Donald G. Newnan, Ph.D., P.E., Civil Engineer, Editor

Written by Six Professors, each with a Ph.D. in Civil Engineering

☆ A detailed description of the examination and suggestions on preparing for it
☆ 195 exam essay and multiple choice problems with a total of 510 individual questions
☆ A complete 24 problem Sample Exam
☆ A detailed Step-by-Step solution for every problem in the book
☆ Printed in two colors

This book may be used alone or in conduction with *Civil Engineering License Review,* 12th edition (Order No. 900). Either way, this book is a complete separate book. The chapter topics match those of *the License Review* book. Here all the problems have been reproduced for each chapter, this time followed by detailed Step-by-Step solutions. Similarly, the 24-Problem Sample Exam (12 essay and 12 multiple-choice problems) is given, followed by Step-by-Step solutions to the exam. Engineers wanting a CE/PE review with problems and solutions, will buy both books. If they only want an elaborate set of exam problems and a sample exam, plus detailed solutions to every problem, then they will purchase this book. 100% Problems and Solutions
ISBN 0-910554-91-9 Hardbound Yellow/Blue 2 color 8-1/2 x 11
Order No. 919 344 Pgs 1995 12th Edition $29.50

Civil Engineering Problem Solving Flowcharts

Jorge L. Rodriguez, P.E., Civil Engineer   Revised Second Edition

89 Flowcharts To Solve Typical Problems On the Civil Engr./Prof Engr. Examination

The flowcharts describe the procedure to follow when solving problems that frequently appear on the CE/PE exam. They combine theory and formulas to show the logical steps in the solution of the problem. Flowcharts are provided for problems in Fluid Mechanics; Hydraulic Machines; Open Channel Flow; Hydrology; Water Supply and Waste Water Engineering; Soils; Foundations; Steel and Reinforced Concrete Design; and other topics. With each flowchart there is an example problem to illustrate the flowcharts Step-by-Step solution procedure. Second Edition. Hardbound Yellow/Blue Cover Full Merchandising Cover 8-1/2 x 11. ISBN 1-57645-012-0
Order No. 120 160 pgs 1997 $27.50

Seismic Design of Buildings and Bridges

Alan Williams, Ph.D., SE, C. Eng.

For Civil and Structural Engineers

✔ Written for Engineers Preparing for the Special Civil Engineer Exam-California, and National Structural Engineering Exam used in 26 states, and the Structural exam used in CA, NV, WA, HI, and ID.

✔ Up-to-date: 1994 Uniform Building Code, and Latest AASHTO, AISC, and SEAOC Standards.

✔ 100 Example Problems of which 50 are multiple-choice problems. Detailed step-by-step solutions for every problem in this book.

✔ 18 Calculator programs to solve the most frequent calculation procedures. Written for HP 28S, these programs present all intermediate stages as well as the solutions.

✔ **8-Page Summary of useful equations,** Use these at test time.

The book has been written to assist candidates preparing for seismic principles examinations. It is a comprehensive guide and reference for self study based on the 1994 edition of the Uniform Building Code. An introductory chapter describes the California Special Civil Engineer and Structural Engineer Exams and the NCEES Structural Examinations. Subsequent chapters cover general seismic principles, static and dynamic lateral force procedures for buildings, seismic design of steel, concrete, wood and masonry structures and seismic design of bridges. Eighteen structural design programs are provided for the HP-28S calculator. One hundred example problems are presented including 50 multiple choice questions. A detailed Step-by-Step solution is provided for all problems. 30%Text   70%Problems and Solutions. Hardbound 8-1/2 x 11 Yellow/Blue

ISBN 0-910544-04-8

Order No. 048   437 Pgs   1995 $41.50

Environmental Engineering Problems and Solutions

Harry S. Harbold, P.E.

More than 100 worked problems for the Environmental Engr. portion of the Professional Engr. exam. Fluid Flow; Water Supply and Treatment; Wastewater Treatment; Sludge Treatment and Disposal; Sanitary Engr. Analysis; Engr. Economy. Appendix. Index. 100% Problems & Solutions. 6x9 Hardbound Yellow/Blue **ISBN 0-910554-79-X**

Order No. 79X   292 Pgs   1/ed $41.50

Structural Engineer Registration: Problems And Solutions

Alan Williams, Ph.D., SE, C.Eng. Registered Structural Engineer-California

A comprehensive guide and reference to assist civil engineers preparing for the Structural Engineer examination. Five hundred twenty three pages of problems with complete step-by-step solutions covering General Structural Principles and Seismic Design; Structural Steel Design; Structural Concrete Design; Structural Timber and Structural Masonry Design. Includes four problems and solutions from California Seismic Principles Exam. Eighteen HP-28S calculator programs. Index.   8-1/2 x 11 Hardbound   Black/Gold Cover   **ISBN 1-57645-016-3**

Order No. 163   565 Pgs   1997 $69.50

Buy any four Civil Engineering PE books in one order and, if requested, we will send *Civil Engineering Rapid Problem Index* as a FREE bonus. Just write "986FREE" on the order.

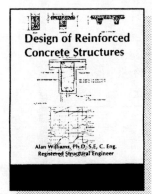

Design of Reinforced Concrete Structures

Alan Williams, Ph.D., SE, C.Eng. Registered Structural Engineer-California

A comprehensive guide and reference to assist civil engineers preparing for the PE Civil and Structural Engineer examinations. Three hundred fifty pages of text and seventy design problems with complete step-by-step solutions covering Materials for reinforced concrete, Limit state principles, Flexure of reinforced concrete beams, Shear and torsion of concrete beams, Bond and anchorage, Design of reinforced concrete columns, Design of reinforced concrete slabs, Footings, Retaining walls, Piled foundations, References, and Structural design programs for the HP-28S. Index.
8-1/2 x 11   Hardbound   Blue/Yellow Cover   **ISBN 1-57645-009-0**

Order No. 090   387 Pgs   1997 $59.50

Civil Engineering Rapid Problem Index
☆☆☆☆
Donald G. Newnan, Ph.D., P.E., Civil Engineer
Essential!
A Key word and Complete problem cross-reference for 8 popular Civil Engineering Review Books: *Civil Engr Lic Review, Civil Engr Lic Prob & Solutions, Design of Reinforced Concrete Structures, Seismic Design of Building and Bridges, Environmental Engr Probs and Soln, Structural Engr Lic Review Prob & Soln, Survey Review for the Civil Engr, Civil Engr Prob Solving Flowcharts.* This is an essential guide for the PE Exam.
Order No. 986 $4.95   1997   32 pages

Order any four Civil Engineering PE review books, and if requested at the time of order, we will send CE Rapid Problem Index free. Ask for "986FREE" with your order.